DESIGN FOR RECYCLABILITY

Michael E. Henstock B.Sc., Ph.D., C.Eng., MIM
The University of Nottingham

DESIGN
FOR
RECYCLABILITY

M. E. Henstock, B.Sc., Ph.D., C.Eng., MIM
The University of Nottingham

Published by THE INSTITUTE OF METALS
on behalf of THE MATERIALS FORUM

THE INSTITUTE OF METALS
1988   8/23/89 97

BOOK 450

published in 1988 by

The Institute of Metals
1 Carlton House Terrace
London SW1Y 5DB

and

North American Publication Center
Old Post Road, Brookfield, Vermont 50536, USA

ISBN 0 901462 46 2

British Library Cataloguing in Publication Data

Henstock, Michael E (Michael, Edward), *1937–*
Design for recyclability.
1. Waste materials. Metals. Recycling
I. Title II. Institute of Metals, *1985–*
III. Materials Forum
669'.028

ISBN 0-901462-46-2

Library of Congress Cataloging in Publication Data

   Applied for

Published by The Institute of Metals
on behalf of The Materials Forum

Typeset by Computerised Typesetting Services Ltd, Redhill, Surrey
Printed in Great Britain at the Alden Press, Oxford

Cover Design from an idea by Luisa S. Henstock

ACKNOWLEDGEMENTS

This project was funded by a grant from The Department of Trade and Industry to The Materials Forum, as a partial response to Recommendation 6 of the House of Commons Trade and Industry Select Committee report on 'The Wealth of Waste', namely:

'The Design Promotion Scheme should be extended to include recycling and reclamation.'

The author acknowledges his indebtedness to his colleagues on the Working Party of the Design for Recyclability Project of The Materials Forum, whose guidance and many constructive criticisms have been invaluable. These colleagues include Mr J. D. Alexander, Mr J. S. O'Neill, Mr F. Roberts, Mr R. Kerry Turner, Mr R. Woodward, Dr W. R. Wilson and Professor J. Nutting, Chairman of the Committee.

He also wishes to thank Mr W. Hague, of The Department of Trade and Industry, for his advice on the preparation of the report.

The work would have been impossible without the generous assistance of many individuals and organisations in the fields of materials supply, product manufacture, and materials reclamation. Many of these supplied material on a confidential basis, and have asked that its source should not be identified.

The opinions expressed are those of the author. In acknowledging the assistance of so many companies and individuals the author must, of course, absolve them from any errors in his interpretation of the material that they have so generously provided.

CONTENTS

PREFACE

The value of the recoverable materials in most products at their disposal is usually small in relation to their initial cost. It is, therefore, difficult to persuade and motivate the manufacturer and first buyer to make any changes to that product which would improve the eventual ease of recovery of its materials of construction. This is especially true if those changes cost money, or have even a small actual or imagined effect on the initial performance or desirability of the product. This report makes it very clear that products can be designed and made in ways which would improve the eventual recoverability of their materials of construction, and this could often be achieved at little or no cost. Design is taken here in its wides sense; the determination and analysis of the need to be fulfilled by the product, conceptual design, selection of materials and its final disposal.

The changing patterns of materials usage also compound the difficulties in relation to recyclability. By their very nature metals are recyclable whereas polymers and ceramics can only be recycled with difficulty if at all. Composite materials and composite structures, where efficient materials use is a prime criterion, almost negate the possibility of recyclability yet the modern tendency is to move towards these advanced materials.

In general it has to be accepted that the financial advantages of easy recyclability are small at the point of manufacture and first sale, but the overall societal value is very great. Equally great are the complexity and sophistication of the measures needed to attain recyclability. This complexity virtually rules out any reasonable possibility of securing action by *legislative* measures, except of the most general sort. Legislation would always lag badly behind the rapidly moving technology, and would be liable to be too broad or too restrictive.

What is called for is the addition of a further requirement to the already formidable list of design considerations, namely final safe disposal and maximum materials recovery, expecially of those chemical elements of limited availability in the earth's crust. Designers, as a group, have always in the past responded responsibly to social needs, and there is no reason to suppose that they will not respond to this call once the need is clear. The problems and the opportunities are set out in a broad way in the pages which follow. Any individual group of products raises specific design problems. It could well be that there is a need for an independent unit to collect and collate experience, disseminate information and tackle problems of design for recyclability.

The second requirement for recycle on disposal is an effective waste and scrap industry. The present infrastructure is an incredible mixture of public and private activity and, within the private sector, a wide range of enterprise size ranging grom large highly organised companies in Steptoe-like organisations. The problems and opportunities are touched on in this report. Their resolution would involve many bodies and several Government departments. More study is needed for, without a suitable waste and scrap infrastructure, better design for recyclability would be wasted. Government initiative, action and encouragement will certainly be needed here. While design aspects of recyclability are only appropriate for rather general measures by Government, there is an exception. This is end use involving very *highly* toxic or physically dangerous chemical mutagenic chemical elements or compounds. These provide two types of problem; firstly, those elements widely used now, such as lead, where better medical knowledge is revealing long-term subtle effects at sustained low concentrations, and secondly elements only now coming into widespread use, such as beryllium and cadmium. It will be argued that ther are many watchdog bodies dealing with this type of problem but to us, approaching the problem from a rather different viewpoint, there seems to be a need for co-ordination and a central body to oversee, advise, warn and possibly even direct.

The underlying aims of this report can also be attained by two other methods in addition to recyclability. They are design for long life (and easy maintenance), and design with remanufacturer in mind. The former is only touched on in his report and the latter is considered at some length. Effective remanufacture and effective maintanance, like effective scrap recovery, calls for an infrastructure quite separate from design, but linked to it.

We commend this report for widespread study in Government, Industry, Academy and especially the Professional Engineering Institutes whose members, as a whole, encompass the design function to which it is primarily addressed.

A. L. Challis
Chairman
Technical Committee
The Materials Forum

J. Nutting, F. Eng.
Chairman
Executive Committee
The Materials Forum

EXECUTIVE SUMMARY AND RECOMMENDATIONS

The investigation examines the factors influencing the recyclability of discarded materials, or of artefacts which contain materials.

In its narrowest sense, recyclability signifies the relative technical ease or feasibility of recovering materials from products which would otherwise enter the waste stream. In the wider sense used here it embraces technical ease of recovery, composition of recovered products, their markets and value, disposal of residuals, and cost of the recovery operation.

In the conditions under which the secondary materials industries operate in Western economies, recovery operations are judged on financial criteria. Thus, the technical ease, and hence the cost, of recovery of a saleable secondary product greatly influences profitability and whether or not a waste is exploited for its materials. Since industry is increasingly obliged to consider the environmentally-sound disposal of its residuals, the ease and cheapness of residuals disposal are financially important.

Recyclability is a relative term because recovery costs are significant only in the context of a given market. Pure reclaimed materials command a premium price and can support high recovery costs. If the reclaimed material is contaminated, or is presumed to be so, it will command only a low price and the scope for complex recovery and separation methods is limited. The overall financial viability of recovery principally depends on the yield and grade of recovered materials, the cost of recovering them, and the cost of residuals disposal. In a free-market economy, markets and prices predominate, and are controlled by the extent to which the recovered materials meet the relevant specifications.

In a general sense, purity, desirability, and value of scrap decline as the reclamation point becomes more remote from the point of materials manufacture. Because of their high monetary value and the fact that secondary can be indistinguishable from primary, metals predominate among the materials that are commonly reclaimed.

Modern materials and manufacturing methods use smaller quantities of materials than hitherto. Nowhere is this more apparent than in the miniaturisation of the electrical and electronics sectors, where minute quantities of precious metals are incorporated in intricate end-products of high added value. Such methods are incompatible with recycling, where recovery operations must be justified by the revenue from saleable products. Another adverse trend is towards coated or other composite materials, especially where the substrate and coating materials comprise a self-contaminating combination.

Recyclability could be improved by design changes in engineering and materials. The former might facilitate contaminant removal by concentrating them in a few well-defined areas of a complex product, and by making them easier to remove. The latter might contribute through the avoidance of coated or composite materials or through the use of materials forming the apex of a hierarchical sequence.

It should not be overlooked that an increased recovery of materials presupposes that the market is capable of absorbing them. Unless it is, recovery serves no useful purpose.

RECOMMENDATIONS

I   Materials identification at scrapyard level should be improved, with a view to preventing the indiscriminate mingling of high-value wrought with low-value cast metals, to the inevitable detriment of the financial value of the former.

II   Design and deployment of alloy systems which, in designated applications, could tolerate tramp elements would provide a sink, and hence an end-use, for contaminated materials.

III   In applications where the recovery of steel contributes substantially to the revenue account the replacement of copper by aluminium wiring should be encouraged. Where copper cannot be replaced, design should endeavour to group copper-containing components together, and preferably to attach them to a valuable component which is usually removed for its materials content.

IV   The widespread replacement of one material by another should not be undertaken without examination of the implications for the secondary materials industries. In particular, the replacement of steel and cast iron by aluminium in the motor vehicle begs the question of whether the market for aluminium castings is capable of absorbing the secondary metal.

V   Replacement of valuable materials, such as steel, by those which presently have no scrap value, e.g. plastics, could make it unattractive to process wastes such as discarded cars. Such possibilities should be considered at the design stage. No material should be introduced without considering its ultimate disposal.

VI   Improvement in scrapyard economics would result from a move to eliminate materials which produce pollution when incinerated. PVC has caused particular interest. The development of low-cost incineration systems, preferably with waste heat recovery, for the disposal of shredder residues at scrapyard level is also desirable.

VII   Formalised collection systems should be considered for low-value discards such as small electronic devices which contain materials of strategic importance.

VIII   Designs should attempt to identify and simplify removal of microprocessors and other components which contain substances, such as semiconductors, whose effects as tramp elements in major recoverable constituents are not yet understood.

IX   Thought should be given at the design stage to the possible effects of a coated or composite material on the composition, and hence the properties, of co-recovered materials.

X   No new substance, especially any known to be toxic, should be used in any application without first considering the hazards of its use in service, problems of recovery at the end of the service life, and the potential for hazardous waste generation.

XI   Design with recyclability in mind would make it easier to establish trust and confidence between scrap processor and scrap buyer. The former would more easily be able to provide a product of consistent quality for which the latter, secure in the knowledge that proper procedures have been observed, would pay a premium price.

Design changes to improve recyclability will not necessarily increase costs, but most seem likely to do so. Thus, unlike changes which reduce costs, they will not come about naturally. They are unlikely to be implemented unless encouraged by legislation, some of which is already in place and which may well produce the stimulus for some change, e.g. by increasing the cost penalties for continuing current disposal practices.

TERMS OF REFERENCE AND GENERAL CONCLUSIONS

The Design for Recyclability project began with the setting up of a Working Party, under the Chairmanship of Professor J. Nutting, with which the Investigator, Dr M. E. Henstock, would work. The Committee comprised:

Mr J. D. Alexander,	Rolls Royce Ltd
Mr S. O'Neill,	BL Ltd
Mr F. Roberts,	*Ex officio* The Materials Forum
Mr R. K. Turner,	University of East Anglia
Mr R. Woodward,	Luxfer Ltd (for the early part of the work), later replaced by
Dr W. R. Wilson,	Alcan International Ltd
Mr A. L. Thomas,	DTI, later replaced by
Mr I. McTaggart,	DTI, later replaced by
Mr W. G. Hague,	DTI

The Committee met on eight occasions: three times in 1985, three times in 1986, and twice in 1987.

TERMS OF REFERENCE:

The level of effort over the two years duration of the project was nominally equivalent to forty working days. Given this constraint, and the complexity of the subject, it was clear that the project had, of necessity, to be selective. The report set out to be illustrative and demonstrative, rather than exhaustive. One of the first tasks of the Committee was, therefore, to identify the areas on which the project would concentrate.

The project was designed to examine the factors influencing the recyclability of discarded materials, or of artefacts which contain materials.

Some materials, notably metals, already have well-established secondary industries. For some metals the secondary industries are capable of producing secondary metal which is indistinguishable, in composition and properties, from primary. Quite clearly, however, the ability to recycle metals to original primary specification may be inhibited if they have been used in association with other metallic or non-metallic materials. Thus, it was appropriate to examine the ways in which design might reduce or eliminate the problems of contamination in reclaimed metals.

By contrast, other materials, such as many plastics, cannot yet be reclaimed economically for use in an application identical to the original. There are chemical changes to be considered, notably the deterioration in properties on successive recycling. The general incompatibility of plastics one with another would require separation processes for mixed fractions far superior to anything currently available. Though there are processes and procedures that can reutilise secondary plastics for the manufacture of items such as dustbins and fence posts, these do not seem to depend upon the initial application to which the primary material has been put. Hence, they fall outside the terms of reference of the study. The recycling of plastics has, therefore, been considered only in outline.

Packaging was specifically excluded from the study because it was considered that, in general, the value of the packaging materials was so small, compared with the retail value of the total product, that it was virtually inconceivable that changes would ever be made for the sake of recyclability. Moreover, the relatively simple nature of most packaging, e.g. a tinplate can, offered less scope for design changes, in materials or methods, than did a relatively complex object, such as a motor car.

After much discussion, it was decided to examine automobiles, gas turbine engines, appliances ('white goods'), electronic goods, and precious metals. During the progress of the work it emerged that it was also desirable to examine hazardous wastes and design for remanufacturing.

Within these limits, there were wide differences in the amount of information that was available in the various sectors of the study. The wealth of data on the materials of construction of the automobile made it the only realistic choice for detailed analysis.

It was felt necessary to transcend the narrow sense of recyclability, i.e. that signifying the relative technical ease or feasibility of recovering materials from products which would otherwise enter the waste stream, and to extend it to the wider sense, where it embraces the technical ease of recovery, composition of recovered products, their markets and value, disposal of residuals, and cost of the recovery operation. This is because of the need of industry to operate within acceptable environmental guidelines and to consider the environmentally-sound disposal of its residuals. Hence, the ease and cheapness of residuals disposal are financially important.

Recyclability is a relative term because recovery costs are significant only in the context of a given market. Pure reclaimed materials command a premium price and can support high recovery costs. If a reclaimed material is contaminated, or is presumed to be so, it will command only a low price and the scope for complex recovery and separation methods is limited. The overall financial viability of recovery depends principally on the yield and grade of recovered materials, the cost of recovering them, and the cost of residuals disposal. In a free-market economy, markets and prices predominate, and are controlled by the extent to which the recovered materials meet the relevant specifications.

In a general sense, the purity, desirability, and value of scrap decline as the reclamation point becomes more remote from the point of materials manufacture. Because of their high monetary value and of the fact that secondary can be indistinguishable from primary, metals predominate among the materials that are commonly reclaimed. It was felt necessary to explain in some detail why the purity of recovered materials is of paramount importance in their acceptability.

Steel was chosen as an illustration, partly because of its predominant position among metals and partly because the value of the steel is often the most important single financial factor in a recovery operation. Steel is especially susceptible to tramp-element contamination. Hence, where its recovery is financially important to a recovery operation, the replacement of harmful contaminants by harmless ones, e.g. copper by aluminium wiring, should be investigated. Where a harmful contaminant ('pernicious contrary') cannot be replaced, design should endeavour to identify and group together those components containing it, and preferably to attach them to a valuable component which is usually removed for its materials content.

There is a clear trend towards the use of smaller quantities of materials per unit of manufacture than hitherto. Nowhere is this more apparent than in the miniaturisation of the electrical and electronics sectors, where tiny quantities of precious metals are incorporated in intricate end-products of high added value. Such methods and products may be incompatible with recycling, where recovery operations must be justified by the revenue from saleable products. Another adverse trend is towards coated or other composite materials; this is especially harmful where the substrate and coating materials comprise a self-contaminating combination.

Recyclability could be improved by design changes in engineering and materials. The former might facilitate contaminant removal by concentrating them in a few well-defined areas of a complex product, and by making them easier to remove. The latter might contribute through the avoidance of coated or composite materials or through the use of materials which could form the apex of an hierarchical sequence of use.

There is a need for better identification of materials at scrapyard level, with a view to preventing the indiscriminate mingling of high-value wrought with low-value cast materials, to the inevitable detriment of the economics. However, given that this is not always possible, design and deployment of alloy systems which could tolerate tramp elements would provide a sink and hence an end-use, for contaminated materials.

The widespread replacement of any material by another has implications for the secondary materials industries and should, ideally, should not be undertaken without examination of those implications. In particular, one may cite changes which could conceivably impair the financial viability of recycling, or which could generate potentially hazardous wastes. An example of the former is the replacement of steel and cast iron by aluminium and plastics in the motor vehicle; this poses questions regarding the size of the market for aluminium castings and its ability to absorb the secondary metal, and of whether the non-metallic waste may be disposed of in an environmentally and financially acceptable manner.

Replacement of valuable materials, such as steel, by those which have no scrap value in the form in which they are *currently* recovered, e.g. mixtures of plastics, could make it unattractive to process wastes such as discarded cars. Such possibilities should be considered at the design stage. No material should be introduced without considering its ultimate disposal. In particular, there should be an attempt to replace substances that cannot be incinerated without production of noxious fumes by substitutes that can safely be incinerated. It is desirable to develop low-cost incineration systems for the disposal of shredder residues at scrapyard level, preferably with waste heat recovery.

The particular problem of formalised collection systems should be considered for low-value discards such as small electronic devices which contain materials of strategic importance. In this context, new materials, such as semiconductors, are being used in applications where they might easily report to a recovered metallic material. Since we do not, at present, have adequate knowledge of the possible metallurgical implications, it would be an advantage if designs were to attempt to identify and simplify removal of microprocessors and other components which contain substances whose effects as tramp elements in major recoverable constituents are not yet understood. The same philosophy extends to the possible effects of a coated or composite material on the composition and hence the properties of co-recovered materials.

In general, no new substance, especially any known to be toxic, should be used in any application without first considering the hazards of its use in service, problems of recovery at the end of the service life, and the potential for hazardous waste generation.

Design changes to improve recyclability will not necessarily increase costs, but most seem likely to do so. Thus, unlike changes which reduce costs, they will not come about naturally. They are unlikely to be implemented unless legislation makes them obligatory. Existing legislation may well produce the stimulus for some change, e.g. by increasing the cost penalties for continuing current disposal practices.

1
INTRODUCTION

The recovery and reutilisation, or recycling, of materials are activities of considerable antiquity and continue to be of great importance to commerce, industry and, to the extent that a recycled waste is avoided litter, to the environment.

Many discarded artefacts that contain potentially valuable materials go unrecovered to the tip or lie unprocessed, disfiguring the landscape. Simultaneously, manufacturing industry purchases and uses virgin (primary) materials, often incurring unnecessarily high costs; these costs can be both financial and, through the mining wastes associated with primary extraction, environmental. To the extent that such primary materials are imported, their use is also a charge on the balance of payments.

The reasons why materials are unrecovered are many and interdependent. Almost without exception, however, a resource will be processed for its materials *if there is sufficient financial incentive to do so under prevailing conditions*. If the resource (residual) remains unprocessed it signifies that the operation is financially unattractive.[1]

Some secondary materials, especially certain metals and alloys, are indistinguishable from primary, with which they are completely interchangeable. However, many discarded consumer products are complex and this may make it difficult and expensive — but not necessarily impossible — to recover from them secondary materials of the consistent quality that renders them acceptable to a potential consumer. Such materials often fail to attain the same specification, and therefore price, as primary. Although there are other factors, it is the unpredictable quality of some secondary materials that may make them unacceptable for some applications. There are, however, usually less-demanding alternative applications for which they may be used. It is the ability of a waste to yield recyclable and saleable materials in a financially and environmentally acceptable manner that is termed 'recyclability'.

There is now an enormous literature on recycling, most of it generated in the decade following the first oil crisis of 1973. Much of it relates to policy, and to possible institutional action to achieve optimal resource use. In Britain and the U.S.A. there still seems to be official ambivalence over whether recycling is more properly a private or a public matter. If it is to be left to private enterprise we should not be surprised if certain low-value wastes remain unassimilated into the materials stream. If, however,

[1] Michael E. Henstock, Realities of Recycling, *Chemistry and Industry*, (17), pp.709–713, Sept. 4 1976.

it is deemed desirable, in the public interest, that such wastes should be reclaimed, then it must be accepted that such an operation must be funded through taxation. An integrated recovery strategy, where materials flows may be projected from the product design stage through to the final reclamation operation may be theoretically attractive but is clearly unworkable in a free-market economy. In practice, all governments interfere in the system to some extent, with taxes on pollution, disposal site licensing etc. Elements of a materials materials policy already exist but they are not integrated and have come into being through piecemeal legislation and through general economic and technological trends.

There is much published material on individual technological innovations in waste processing. Very little interest has, however, been shown in an analysis of the factors that contribute to recyclability.

The improvement of recyclability poses two main problems:

(a) Separation should be made more complete, i.e. the recovered fractions should be purer, and hence more valuable.

(b) Separation should be made easier, and hence cheaper.

These aims might be realised, at the design stage or later, in the following ways:

(a) Changes in engineering design

- Mechanical disassembly might be simplified.
- There might be an effort to avoid self-contaminating combinations of materials.

(b) Changes in materials

- Materials might be standardised.
- Materials might be identifiable.
- Harmless materials might replace deleterious ones.

(c) Changes in reclamation techniques.

Areas (a) and (b) are those which are relevant to this study. Area (c) will be discussed only briefly.

2
RECYCLABILITY AND DISPOSABILITY

2.1 INTRODUCTION

The recovery and utilisation of discarded materials are ancient and universal activities whose economic importance is often unrecognised.

Recycling is a valuable element in the rational use of scarce or potentially scarce materials.[1] The general advantages of recycling *per se* have been amply reviewed in the literature, and need not be repeated here. Its productivity and benefits can almost always be increased, however, by appropriate technological and institutional intervention.[2]

The term 'recycling' refers, in the strict sense, to the reassimilation of a material into the raw materials stream in a manner such that it may be used for a purpose identical or similar to that of its first use. Recycling must be distinguished from 'reuse', which properly refers to the reutilisation of an object, such as a refillable bottle. It must also be distinguished from the several activities that provide a manufacturer with a secondary material for reassimilation; these include 'recovery' or 'reclamation'. Only when a material has been recovered or reclaimed from a waste stream is it available for recycling. Recovery or reclamation must therefore precede recycling.

In its narrowest sense, the term 'recyclability' signifies the relative technical ease or feasibility of reintroducing a particular material, recovered from products that would otherwise enter the post-consumer waste stream, into the raw materials supply. This implies *materials* recovery as opposed to energy extraction.

In a wider sense, however, 'recyclability' includes:

(a) Ease of recovery, or of separation from wastes.
(b) Specification and acceptability of recovered products.
(c) Markets for recovered products.
(d) Disposal of residuals, left after recovery of valuables.
(e) Cost of recovery of valuables, and of disposal of residuals.

The maximum tonnage of a material that can be recovered at any date is a function of the quantity put into service one average product lifetime

[1] NATO Science Committee Study Group, *Rational Use of Scarce Metals*, p.33 *et seq.* Brussels, NATO Scientific Affairs Division, 1976.
[2] Charles G. Gunnerson, *Research and Development in Integrated Resource Recovery:* (An Interim Technical Assessment.) United Nations Development Programme Project GLO/80/004, July 31 1984.

earlier. For copper, such a lifetime might be thirty years; in 1953, world production of refined copper was $3 \cdot 4 \times 10^6$ tonnes. Even if this could be recovered completely — a manifest impossibility — it could provide no more than 35% of the 1985 production of $9 \cdot 7 \times 10^6$ tonnes of refined metal.[3]

Although continuous technological progress improves the efficiency of materials use, consumption of most raw materials continues to rise. Therefore, recycling can provide relief only in the short term. In many cases, materials are not recovered because it is not financially attractive to do so. To maximise the contribution of recycling, it is clearly important that it be made as easy as possible to recover materials from the objects in which they have been used.

2.2. DEPENDENCE OF PROPERTIES ON THE PURITY OF MATERIALS

Materials are substances for making useful objects. They must possess a satisfactory combination of properties such as corrosion resistance, ductility, electrical conductivity, impact resistance and tensile strength, and this combination must be attained at minimum total cost of the object in place. Total cost depends not only on the cost of the material itself but also on the cost of conversion, on the materials wastage rate, and on the value of the residuals. This last will be positive in the case of scrap metals and negative in a case such as mixed non-metallic *detritus* from a scrap shredder, where the residuals currently have no monetary value and must instead incur disposal costs.

Many properties of materials depend on composition. For example, the electrical conductivity of copper is impaired by small amounts of contamination (known as 'tramp elements'), especially by metals of high valency; steel machinability is improved by controlled additions of sulphur and manganese; corrosion resistance of zinc-based die castings is adversely affected by traces of cadmium, lead or tin, and hot workability of steel is dependent on limiting copper and tin to approximately $0 \cdot 2\%$ and $0 \cdot 06\%$ respectively.

2.3 ORIGINS OF CONTAMINATION IN SECONDARY MATERIALS

Primary materials specifications are usually controlled by the producer to meet industry-wide standards and may present no problems, since most smelters apply standard refining procedures to ores whose composition changes only over a period of years. By contrast, secondary materials arise from a wide range of sources, each with a different potential for contamination. In a general sense, the more remote the point of discard of the scrap relative to its point of origin the greater is its potential for contamination. Secondary material is a recirculating load which picks up further contamination on each successive remanufacture, raising the

---

[3] *Metallstatistik 1975–1985*, 73rd Edition, Frankfurt am Main, Metallgesellschaft Aktiengesellschaft, 1986.

cumulative impurity level of new material stock to a point at which the material may become unserviceable, and hence unacceptable, in a particular application. This constraint effectively limits the proportion of scrap material of given composition that may be incorporated in a given furnace heat in the production of material of given specification.[4] There is evidence that copper and tin levels in U.K. steel showed a marked rise during the Great War, almost certainly owing to the increased use of scrap, and have never reverted to pre-1914 levels. The situation can only worsen when large quantities of post-consumer tinplate can scrap or of copper-containing, corrosion-resistant pipe steel enter the ferrous scrap system.

The nature and extent of contamination depend upon the initial application. Alloying, fabrication, joining, finishing and service each provides contact with and possible contamination by other materials. For example, scrap steel from shipbreaking is of heavy gauge, with a high volume to surface area ratio, and is likely to be relatively free from metallic and non-metallic impurities. By contrast, used motor cars are much less desirable since a baled vehicle hulk will contain all the contaminants that have provided insufficient financial incentive for their removal. These can vary from whole tyres and seats to lead-based panel filling and copper wiring. A general trend is observable towards the use of smaller quantities of metals in consumer goods, and their replacement by polymers and other non-metallics. A further trend is toward the attainment of required properties by the use of complex materials or mixtures of materials; this is evident in the use of coatings to confer specific properties on surfaces, and in the use of composites, such as fibre-reinforced materials. Such measures, often adopted to reduce first cost, almost always have an adverse effect on the value of recoverable materials. Taken to its logical conclusion, this trend could lead towards the virtual cessation of the recovery of secondary materials from post-consumer scrap and a corresponding increase in the amount of material discarded in irrecoverable form.

2.4 INTERNAL AND EXTERNAL COSTS

Several other factors, some of which are non-quantifiable, must be included in any analysis of the desirability of recycling. Producers of secondary materials from scrap and the users of those secondary materials normally concern themselves only with their own individual costs, which are easy to identify in the appropriate balance sheet.[5]

However, there are other, so-called 'external' costs, which relate to the use of common property resources, such as air, water or land. These resources are used for residuals disposal, but which the system currently

[4] E. J. Duckett, The influence of tin content on the reuse of magnetic metals recovered from municipal solid waste, *Resource Recovery and Conservation*, **3**, (3/4), pp.301–328, Nov. 1977.

[5] The author has preferred to use the term 'individual costs' in order to emphasise that these considerations apply equally to all materials-consuming and residue-generating enterprises, whether privately or publicly-owned.

employed fails to allocate efficiently among all users. At present, the cost of disposal in common property resources is borne largely by the general public in the form of impaired environmental quality. Thus, the optimum recycling level for an individual enterprise does not necessarily correspond with the social optimum.[6]

As an example, individual considerations might dictate that it is not worthwhile to recover more than the first — and cheapest to recover — 75% of a particular discarded material simply because the higher-value markets for it absorb only that proportion and increased recovery would not, therefore, be rewarded *pro rata*. Similarly, individual considerations might suggest that it is cheaper to landfill certain residues rather than incinerate them for heat recovery, since the latter course may involve unacceptable costs in gas-cleaning. Since the U.K. scrap-processing industry is, in principle, independent of the materials-consuming industries and of those responsible for energy supply, there has been no effort to optimise the system nationally.

This is not to suggest that the reclamation industries are in any way less responsible than other industries, most of which apply the same criteria. The argument is developed simply to emphasise that there is no apparent connection between the designers and manufacturers of goods, the reclamation industry, the users of reclaimed materials, the suppliers of energy, and the waste-disposal authorities. Except in cases of exceptionally hazardous or toxic wastes the degree to which materials are reclaimed is, in practice, directly or indirectly driven by market forces alone.

2.5 DISPOSABILITY

To the extent that the disposal of residuals presents a problem, it may be termed 'disposability'.[7] Here too, however, 'cheaper' usually relates to individual rather than to global costs, and disposal as usually practised may externalise some of the relevant costs. Since 'recyclability' depends on costs, some of which are necessarily associated with disposal, the question of recyclability cannot be separated from the question of disposal of the residuals from a recovery operation.

---

[6] Walter O. Spofford Jr., Solid residuals management: some economic considerations, *Nat. Resourc. J.*, **11**, (3), pp.561–589, July 1971.
[7] Residuals may be defined as those products that it is cheaper to discard than to utilise.

3
FINANCIAL AND ECONOMIC ASPECTS OF MANUFACTURE, DISTRIBUTION AND DISCARD OF GOODS

3.1 INTRODUCTION

Recyclability is a relative term because ease of recovery from a given product and the consequent profitability of the operation depend on a number of interdependent factors. In some cases where materials are not recovered it is a simple matter to identify a single, major reason. In other cases, the reason may be a complex interaction of several factors. Though a small change in any of them may make recovery more attractive, complementary improvements in more than one area may be necessary to make a given recovery economically viable.

The principal operations in recycling are:

(a) collection
(b) transport
(c) separation
(d) aggregation
(e) reutilisation, including marketing

3.2 CONDITIONS OF PRODUCT DISCARD

If materials recovery is the aim, product discard must be clearly distinguished from product disposal. The terms should not be used indiscriminately. The initial need is to collect discarded products so that they may be processed to yield recyclable materials. Only residuals which offer no financial return under current or foreseeable conditions should be sent for disposal.

Collection is the key to many recovery operations. Many discarded products are financially attractive, and are salvaged, officially or unofficially, for spare parts or for non-ferrous metals. Others, which are less attractive because they contain smaller quantities of less-desirable materials, may still be salvaged and, by means of intensive hand-labour, yield saleable materials if:

(a) labour is cheap and readily available,
(b) overheads are low, or
(c) salvagers can operate unofficially, thus avoiding taxes and environmental constraints.

7

Salvage operations which fall into any or all of these categories may profitably collect discarded products and process them for saleable materials. However, they will collect only when conditions make it attractive to them; in this they differ from the municipalities, who have a statutory obligation to collect refuse and other discarded objects. They can be selective in a way that the municipalities cannot.

The dividing line between attractive and unattractive waste arisings is not a rigid one; it can alter in the short term. When changes in the markets for recovered materials make a product unattractive to salvagers, discarded products become a disposal problem and materials are lost. However, if products could be made more amenable to materials recovery by mechanised, or at least standardised methods — i.e. made more recyclable — reclaimed materials would be cheaper. It follows that there would be a greater, and possibly a more continuous demand for them, thus stimulating greater investment in a particular recovery operation.

There can be no question that many products are capable of yielding recyclable and saleable materials. The reason why, so often, they do not is that it is financially unattractive to do so. If the cost of collection, transport and separation exceeds the value of the reclaimed materials, it follows that the discarded products will eventually become a disposal problem.

Many discarded products contain only small quantities of recoverable materials, and their unit recoverable values are low. If the products are themselves physically small, eg. pocket calculators, unserviceable units may conveniently, and at no cost, be discarded in the refuse stream, and therefore lost. Larger objects, such as washing machines, contain greater quantities of materials, and may repay transport costs provided that materials recovery from them is not unreasonably complex or expensive.

Although collection may present significant obstacles to recovery, it cannot be divorced from the technical problems of separation. If separation can be made cheaper, collection for separation must become more attractive.

3.3 COLLECTION SYSTEMS

It is not intended to analyse the abundant literature on the theory and practice of collection systems. As examples, the collection of used tinplate cans is possible via the refuse stream. Larger items, such as refrigerators, might best be dealt with through the distribution network and a system of taxes and incentives. Bottle banks and reverse-vending machines exemplify only two possible approaches to the problem.

The financial viability of collection systems depends upon current conditions in the markets for secondary materials. The collection problem assumes particular importance in low-value, small-bulk articles that are easily disposed of via the refuse stream, and are therefore lost. Such a case is the electronic, hand-held calculator, each of which contains only minute amounts of precious metals. However, they are sold in large numbers and are now so cheap that they are seldom worth repairing, and may be considered disposable. The question then arises of whether, in the national interest, the prospective loss of precious metals justifies the setting up of some type of collection system.

3.4 AVAILABILITY OF RECOVERY TECHNOLOGY

Contamination of recovered materials may be avoided by efficient segregation. Hand stripping is, in most cases, very efficient in the dis-assembly of complex structures, but marginal labour costs can easily exceed marginal revenue. It cannot, moreover, deal with combinations, such as alloys or surface-coated materials.

Less efficient but, in the industrialised, high-wage economies, hitherto less expensive than hand-stripping, have been mechanical separation methods, such as shredding. These are energy intensive. Hand-stripping a motor car may require about 10 MJ whereas a 5000 horsepower shredder, processing one such vehicle every 30 seconds, requires 112 MJ for the shredder alone, taking no account of energy consumed elsewhere in the plant. Since, in the less-developed countries (LDC), labour rates can be as low as $1/day, it is not usually worthwhile installing capital-intensive equipment there to perform tasks that may be carried out adequately by the available cheap labour.

Where capital is available the scrap industries have access to a wide range of separation systems, many of them developed for treatment of arisings such as the non-ferrous fraction from auto-shredders. Such systems do not, however, generally yield products that are capable of immediate reassimilation into the materials stream *in the same form as their original application*. Refining is often necessary. More often, they may be reassimilated in degraded form, where they perform the valuable functions of supplementing or replacing virgin materials. (See Chapter Twelve)

There is a need, in the scrap industries, for efficient, reliable and cheap methods of identification of materials. In many cases, identification is still made by a knowledge of the application in which certain materials are to be found, or by slow, traditional methods which rely on acquired skills.

3.5 CAPACITY OF THE CONSUMING INDUSTRIES

Demand for secondary materials is a function of their utility. Thus, for example, production of wrought aluminium alloys requires high-grade new material of known specification that is unlikely to be met by most of the scrap grades arising from, for example, scrap shredders.

The uses for highly-contaminated secondary products are limited, since they lack the flexibility of high-grade materials. Hence, they command a low price and the markets for them are volatile.

The ability of the consuming industries to absorb recovered materials is dealt with in Chapter Twelve.

3.6 DESIGN OF POTENTIAL RESOURCE

The principal concern of designers is that products shall fulfil the desired function at minimum cost. This has traditionally been achieved through economies of scale. Now, however, components are being redesigned to reduce materials consumption, or to facilitate materials substitution by less expensive alternatives. These actions are very likely to impair recyclability. For example a solid brass/copper car radiator removed from a hulk might simply be remelted. A redesigned equivalent, however, may include aluminium, copper, polymers, and steel; hence, it must first be dismantled, incurring cost, and may well provide material of lower monetary value per unit.

Efficient separation of the materials of any product cannot be performed without cost. So far as the individual scrap processor and scrap consumer are concerned the pertinent question is whether the sum of all the costs exceeds the sum of all the revenues. Both processor and consumer are, however, normally concerned only with individual costs.

On a wider scale, the important question is whether the sum of the individual and external costs associated with the use of a given quantity of material is less if that material is processed to obtain recyclable material than if the object is disposed of as residuals after the end of its useful life.

The reason why discarded materials are not reabsorbed into the materials stream is generally because it has not been financially worthwhile so to do. A reduction of recovery costs might result from engineering or materials design changes to enhance recyclability, viz:

(a) Easing of mechanical disassembly of complex products.
(b) Increasing materials standardisation.
(c) Specific chemical compositions of materials might be made easily identifiable via tracing mechanisms.
(d) Improved chemical or physical separability.
(e) Improved separation techniques for complex materials.
(f) The use of combinations of materials that are not mutually incompatible.

Since most products are designed to minimise initial cost it follows that design changes would almost certainly raise the cost. Most markets are competitive and there is no evidence that the consumer would willingly pay a premium for so remote a benefit as improved recyclability. Since any manufacturer designing to a criterion of recyclability would probably be at a competitive disadvantage it is intrinsically unlikely that this will occur unless legislation is introduced. It is also evident that customer preferences in style and finish may be inconsistent with maximum recyclability, and no early reversal of this trend seems likely.

It has been suggested that design engineers and design executives should begin treating recyclability as on a par with the appearance and durability of their products. Industry, however, does not and, arguably, cannot take this view. For example, conversations with, for example, vehicle and appliance manufacturers both in the U.K. and the U.S.A. make it clear that, although they are conscious of the problems of recycling, concepts of recyclability do not even enter into their list of priorities, which must be based on cost and competitiveness. Hence, styling is of considerable concern to them, irrespective of the possibly adverse effects that changes made to miniaturise and streamline might have on the ease of disassembly.

Profitable dismantling of everyday items to yield usable grades of material is often complicated by constructional features that have been adopted to reduce the cost of manufacture. Composite materials or coated components bring a material into intimate and often permanent association with another which may be damaging to its properties. Some components, especially in electronic equipment, have been reduced to very small sizes, such that recovery is intricate and the yield small. Progressive replacement of materials that possess scrap value by those for which no well-defined secondary market exists, eg. substitution of polymers for metals, may be shown to have an adverse effect on the revenues attainable from processing arisings such as car hulks. Finally, the sheer complexity of construction of items such as an oil-filled submarine cable, which may contain cadmium, copper, jute, lead, oil, paper, polythene and steel may, under all but the most favourable market conditions, defeat efforts to reclaim the valuable materials.

Except in the last of these examples, where reliability and longevity may transcend first cost, there is little doubt that a desire to reduce costs has been responsible for changes in materials and in manufacturing methods. On average, some 40% of the cost of typical products is accounted for by raw materials and this, with increasing automation in manufacture, seems likely to rise rather than fall.

There is, however, now some evidence of awareness among engineers of the importance of considering recycling at an early stage of design.[1] In some cases, such as aviation, considerations of safety or performance must be paramount and there is little or no possibility that recyclability can play any part in design decisions. In others, however, there are reasonable alternative materials whose use could improve the scope for recycling.

[1] LaVerne Leonard, Specifying Materials for Recycling, *ME*, Sept. 1985, pp.47–50

4

THE PLACE OF SCRAP IN THE MATERIALS SUPPLY SYSTEM

4.1 INTRODUCTION

The term 'scrap' has several meanings, some of them pejorative. In the present context, it relates to material that is one of the following:

(a) a by-product of the materials-manufacturing process, (known as Home Scrap),

(b) a by-product of the materials conversion process (known as New Scrap or Prompt Industrial Scrap), or

(c) a residue, in the form of a worn-out or otherwise unserviceable final product (known as Old Scrap or Obsolete Scrap).

'Home', 'new' and 'old' scrap are each known under a variety of other different names.[1]

In principle, scrap arising from any of these three categories can be reassimilated into the materials stream. In practice there are profound differences in the purity, and hence acceptability of the types and in their relative desirability and monetary value.

An understanding of these differences is crucial to the question of recyclability.

The importance of scrap in the metal-producing industries, and the particular problems that arise in the use of metals recovered from post-consumer goods, may be illustrated by a detailed consideration of scrap use in steel which is in tonnage terms the most widely used metallic material. Though described in the context of steel, most or all of the problems may also be found, in one form or another, in the recycling of other metals and non-metals.

4.2 SCRAP STEEL

The recycling of steel is an integral part of the ferrous materials supply system, and can be discussed only within the framework of that system.

To introduce perspective, it is relevant to examine some statistics. In 1984, the U.K. steel industry used $7 \cdot 86 \times 10^6$ tonne of scrap in the production of $15 \cdot 12 \times 10^6$ tonne of steel. Of this scrap, $4 \cdot 29 \times 10^6$ tonne, worth

---

[1] Michael B. Bever, The Recycling of Metals: I.Ferrous Metals, Conservation and Recycling, Vol.1, No.1, pp.55–69, 1976.

approximately £250 million, was purchased from scrap merchants or directly from scrap producers.[2]

The extent to which secondary metal is used in the iron and steel industries of the world depends, in a complex manner, on a number of factors.[3] The two main influences are:

(a) The availability and cost of competitive raw materials, i.e. direct-reduced ores, hot metal,[4] or pig iron, relative to scrap.
(b) The type of steelmaking furnace employed.

The factors affecting the choice of iron source are many and inter-dependent. They may be technological, including heat balance, the quality of the input and of the desired product, the type of energy source available, and the type of furnace employed. Economic factors include the cost and availability of competing sources of iron units, of other raw materials, and of capital.

4.2.1 Types and origins of scrap

Residuals of any material occur in its production and in the processing, use and discard phases of industrial activity. They may be classified according to origin as follows:

4.2.1.1 *Home scrap*

Materials production is always accompanied by the generation of scrap; in steelmaking, it arises from activities such as the cropping of ingots and the edging of sheets. This material, known as 'home scrap', is readily accessible, of known composition, and effectively free of charge. Thus it is almost always completely recycled. The quantity of home scrap generated is a function of the amount of metal produced. That quantity does not depend on price, i.e. home scrap is price-inelastic.[5]

4.2.1.2 *Prompt industrial scrap*

The conversion of materials produces scrap called 'prompt industrial' or 'new' scrap. It is generated during the manufacture of products; new steel scrap is sold in the form of stampings and clippings, compressed into bundles. Although of slightly greater potential for contamination than is home scrap, it is a desirable grade. Like home scrap, new scrap is essentially price-inelastic.

4.2.1.3 *Obsolete scrap*

Obsolete goods containing recoverable materials are discarded by many different classes of user, including utilities, transport, companies and households. The saleable portion of such goods is called 'old',

---

[2] A. R. Tron, *Contaminants in Ferrous Scrap*, Recycling Advisory Unit, Warren Spring Laboratory, Nov.1986.
[3] Michael E. Henstock, The Occurrence, Utilization and Future Prospects for Ferrous Scrap, *Proc. Seminar on Scrap Recycling, Iron and Steel Institute of Japan*, Tokyo, April 10 1981.
[4] The term used in iron and steelmaking to denote blast furnace iron, or molten pig iron.
[5] Price elasticity is defined as the percentage change in supply for a 1% change in price.

'obsolete' or 'post-consumer' scrap. Quality ranges from low to high, depending on size, density and purity; in steel, the later relates to its freedom from non-ferrous metallic and non-metallic impurities. By its nature, obsolete scrap has been exposed to a greater variety of contamination than have home scrap or new scrap. For example, steel scrap from demolished buildings or from shipbreaking is relatively free from metallic contamination and its high density means a high yield on remelting. It may require very little processing to prepare it for furnace use. Other grades, including appliance, automotive or tinplate scrap, are much less desirable. In a general sense, the more distant the point of discard of the scrap, relative to the point of materials manufacture, the greater is the potential for contamination. Alloying, fabrication, joining, finishing and use phases each provides contact and possible permanent association with contaminants. Obsolete scrap can exhibit high and often unpredictable levels of residual element (tramp) contamination. For this reason its use is strongly linked with general economic activity; in a recession, when superior grades of scrap are likely to be in surplus at depressed prices, the market for obsolete scrap may almost disappear, no matter how low the price. By its nature, obsolete scrap has some price elasticity thus, higher prices at the steelworks will generally result in more intensive efforts to locate and transport such scrap to the steelmaking furnaces.

4.2.1.4 *Relative potential of scrap types as a source of iron units*
Steel can be produced from:

(a) liquid iron, made from ore, or
(b) scrap.

The production of primary iron is dependent on the availability of capital for large, integrated iron and steel plants. These are inherently inflexible; they cannot be built to respond to urgent need, and they cannot economically be closed down should there be a temporary recession and a slack market.

By comparison, steel scrap melting facilities can be built quickly and cheaply. In particular, the electric arc furnace provides the potential for rapid increase of steel production capacity.

The production of home scrap and of prompt industrial scrap is, however, tied to primary production and cannot be increased independently of general economic activity which, while generating scrap, also increases the demand for raw materials and hence for scrap. Home scrap, moreover, is always consumed by the producing industry, and is hardly ever available on the open market.

Hence the only way in which steel production can be increased in the short term is by the use of obsolete scrap.

15

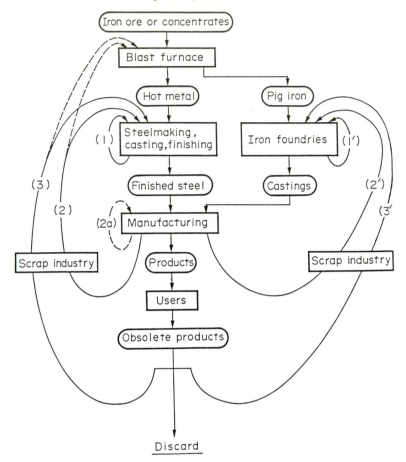

Fig. 1   After Bever

4.3   STEELMAKING PROCESSES

Many of the economic and certain of the technological influences of steelmaking fluctuate even in the short term; they cannot be termed constant factors. It is therefore appropriate to consider, first, those that, even if not strictly fixed, are least capable of rapid change. These comprise the capital-intensive fixed equipment, principally the steelmaking furnaces.

In the developed world, three types of steelmaking furnace are in common use:

(a) The open-hearth furnace (OHF)
(b) The basic oxygen furnace (BOF)
(c) The electric arc furnace (EAF)

4.3.1 The open-hearth process

The basic open-hearth furnace, which produces steel by the oxidation of impurities in a furnace heated by gas or oil, is in decline. It is obsolete in the U.K. and Japan, but is still used in the U.S.A. and elsewhere.

Because its heat supply is under external control, the open-hearth furnace is much more flexible, in its capacity for consuming scrap, than is the basic oxygen furnace, which relies on the exothermic oxidation of impurities in its charge. In fact, from the standpoint of the thermal balance, the open-hearth furnace can use charges wholly of scrap or of hot metal, but these extremes are undesirable for economic reasons. The cycle length, some 5–10 h, facilitates good control of impurity removal and hence of steel composition. Still another advantage so far as scrap consumption is concerned is that the size and shape of the furnace permit it to accept large pieces of scrap, e.g. baled motor vehicles, which are dimensionally unacceptable in competing furnace types.

Typically, scrap comprises 40% of charges in hot metal plants and 85% in cold melt plants. In 1983, the average scrap rate for all open-hearth furnaces in the U.S.A. was 45%.[6]

The general decline in open-hearth steelmaking is largely a function of its economic disadvantages *vis-à-vis* competing processes. As recently as 1955, the open-hearth process produced up to 90% of U.S.A. steel output. In or around that year, however, the basic oxygen furnace was introduced. This, with its principal advantage of short cycle times and, therefore, high productivity, grew rapidly in popularity, and overtook open-hearth furnace tonnage in 1969.

4.3.2 The oxygen converter

In the basic oxygen furnace, essentially a top-blown converter of which several variants exist, heat is generated internally by the oxidation of the impurities in the charge. The heat balance is therefore determined by the temperatures of the hot metal and of the scrap, their relative amounts, and their respective purities. In a given shop, hot metal temperature and composition are fairly constant, and it is common practice to fix, at least in the short term, the relative proportions of hot metal and scrap and to balance heat requirements with iron ore. Scrap is added primarily as a coolant rather than as a source of iron units. In practice the charge includes 25–30% scrap, although by preheating the scrap or by fuel injection it is possible to increase this level. Since in a steelworks that produces its raw metal as single ingots (rather than as continuously-cast stock) about 30% of the metal poured will become home scrap, the basic oxygen furnace can be supplied with scrap entirely from this source, and none need be acquired externally. Thus, the replacement of the open-hearth furnace by the basic oxygen furnace has been accompanied by a reduced demand for scrap.

[6] *Facts, 1983 Yearbook,* Washington, D.C., Institute of Scrap Iron and Steel, Inc., 41st Edition, 1984.

A general move towards continuous casting of steel should be noted with, typically, a consequent reduction in home scrap generation from 30%, with the traditional ingot route, to 5%.[7] Thus, there should in principle be a shortfall of scrap for the converters. The integrated steelworks could meet such a shortfall with the blast-furnace hot metal made available by the reduced demands of continuous casting, relative to the traditional ingot route; there are, however, implications for the thermal balance.

4.3.3 The electric arc furnace

The electric arc furnace was originally used in steelmaking to produce alloy grades. Latterly, however, probably in response to the ready availability of cheap steel scrap which was released as the open-hearth furnace yielded to the basic oxygen furnace, it has been used as the basis of so-called 'minimills'. These are steelmaking facilities without blast furnaces; instead, they operate entirely, or almost so, on scrap. The mills are designed to produce not high-quality steel but products such as reinforcing bars for concrete, often for a local market.[8]

In the period 1976–1985, electric arc furnace production of steel rose from 19·2 to 33·2% of total output in the U.S.A. In the same period, basic oxygen furnace production fell from 62·5 to 59·5%, whilst output from the open-hearth furnace declined from 18·3 to 7·3%.[9] Output of the minimills has risen from 2·0% in 1960 to 17·6% of total U.S. steel production in 1985.[10]

4.4 REQUIREMENTS OF THE SCRAP STEEL CONSUMER

The aim of the steelmaker is to obtain maximum output at minimum cost. The steelmaker's criteria for scrap availability may be divided into physical and chemical factors.

4.4.1 Physical factors

Scrap processors and users attempt to achieve maximum scrap density, which reduces transport charges and surface oxidation, shortens the charging and melting times,[11, 12] and is beneficial in continuous melting, refractory protection and avoidance of electrode breakage.

---

[7] Henstock, The Occurrence, Utilization and Future Prospects for Ferrous Scrap, *loc. cit.*

[8] Donald F. Barnett and Robert W. Crandall, *Up from the Ashes: The Rise of the Steel Minimill in The United States,* Washington, D.C., The Brookings Institution, 1986.

[9] Staff, Institute of Scrap Iron and Steel, Inc., Facts, 1985 *(43rd Edition),* Washington, D.C., Institute of Scrap Iron and Steel, Inc., 1986.

[10] Barnett and Crandall, *op. cit.*, p.12.

[11] Ernst Amelung, Physical and chemical requirements for scrap deliveries to steel mills, *Meeting of the Bureau International de la Récuperation,* Scrap Iron and Steel Division, Madrid, May, 1976.

[12] *Iron and steel scrap consumption problems,* Washington, D.C., US Department of Commerce, Business and Defense Services Administration, March, 1966.

4.4.2 Chemical factors

Chemical factors are considerably more complex. The specification for any material is related to the conditions of service for that material and to the production processes that it undergoes; such a specification calls for impurities to lie within bands that are more or less narrow, according to circumstances.

The discussions that ensue are not intended to discuss, however briefly, the entire range of residuals in steel. Detailed treatments of these are readily available.[13] They serve only to illustrate the general problems that may arise from the use of scrap. The capacity of the consuming industries to absorb contaminated scrap is discussed in Chapter Twelve.

The residual elements; commonly found in steel are those which oxidise less readily than iron, or that form oxides which are insoluble in the slag. Any elements not intentionally charged to a steel melting furnace are considered a residual or tramp element. Depending on their nature and concentration, they may be beneficial, harmless, or deleterious in the metal or to the refractory.[14,15] Certain elements are removed during steelmaking and others may be counteracted by additions; however, the deleterious effects of others may be minimised only through dilution. Table 4.1 indicates the general destination of tramp elements.

Dilution of melts with a high concentration of tramp elements may necessitate so much pouring-off to make room in the furnace for additional diluting material that the additional cost in time, labour and

Table 4.1
Effect of steelmaking operations on elements present in the charge (after Boulger, *loc. cit.*)

Predominantly recovered in the steel	Partially recovered in the steel	Almost entirely eliminated from the steel
Antimony	Carbon	Aluminium
Arsenic	Chromium	Calcium
Cobalt	Hydrogen	Magnesium
Copper	Lead	Silicon
Molybendum	Manganese	Titanium
Nickel	Nitrogen	Zinc
Tin	Phosphorus	Zirconium
Tungsten	Sulphur	
	Vanadium	

[13] A. Kelly, D. W. Pashley, E. D. Hondros and C. Lea (Editors), *Residuals, Additives and Materials Properties*, London, The Royal Society, 1980.
[14] Francis W. Boulger, Effects of residual elements in steel, Joint Technical Session: Minor elements in steel, *A.I.M.E. Electric Furnace Proc.*, **20**, 1963.
[15] C. B. Jacobs, J. F. Elliott and M. Tenenbaum, *Blast Furnace and Steel Plant*, **42**, (6), June 1954.

materials does not justify the exercise. A contaminated melt may need to be taken to the scrap pile and/or held to supply a future order with less rigid specifications.

Although the elements retained in the steel are potentially those most damaging to the refractory others, such as lead, can damage it by physical penetration.

Some types of steel product, such as the bars used to reinforce concrete, have almost no limitations on; non-ferrous elements, or at least none that could be reached accidentally through tramp elements. However, for the hot rolling of very thin sheets for complex fabrication, such as car body panels, residual levels are very critical indeed. Since many impurities are wholly transferred from the metallic input to the steel output, the technical feasibility of the inclusion of steel recovered from products, such as vehicles, in a steelmaking heat depends on the balance of three factors:

    (a) the desired product mix and the related specification for each product,

    (b) the level of impurity removal before melting, and

    (c) the dilution in steelmaking.

4.4.3   Effects of residuals in steel

4.4.3.1 *Copper*

It is generally accepted that copper, although sometimes added to confer corrosion resistance, is the most troublesome of the residual metallic elements found in scrap steel, since it usually occurs in larger concentrations than do other contaminants and it is not removed during steelmaking.[16, 17] Many steelmakers confirm that, for all practical purposes, copper nears its maximum permitted level in a particular furnace heat more often than does any other element. Since iron ores contain a maximum of 0·1% Cu, and generally much less, the problem must arise largely from scrap.[18]

The reason why most steelmakers place an upper limit on copper lies in the effects that it has on surface hot shortness. The presence of 0·1% Cu raises the hardness of annealed steels only slightly, but at above 0·2% it increases strength and reduces the elongation and the work hardening modulus. It has been claimed that hot rolling is a problem only when copper exceeds 0·5%[19] but other data suggest that this figure is far too high.[20, 21, 22] Heating in sulphur-bearing furnace atmospheres accentuates

[16] Richard D. Burlingame, A scrap gap in the space age, *Scrap Age*, June 1970.

[17] Boulger, *loc. cit.*

[18] Jacobs *et al.*, *loc. cit.*

[19]Boulger, *loc. cit.*

[20] B. B. Hundy, Residual elements in carbon steels, *Metallurgical Developments in Carbon Steels,* Iron and Steel Inst. Special Report, No.81, 1963.

[21] D. A. Melford, Surface hot shortness in mild steel, *J. Iron Steel Inst.*, **200**, (4), April, 1962.

[22] D. A. Melford, Influence of antimony and arsenic on surface hot-shortness in copper-containing mild steels, *J. Iron Steel Inst.*, **204**, (5), May 1966.

the hot brittleness caused by copper; nickel, also, can be injurious in such an environment, owing to the low melting point of nickel sulphide.

During surface oxidation of steel, the loss of iron through scaling is accompanied by a surface build-up of elements that are nobler than iron. Enrichment in antimony, arsenic, copper, nickel and tin can thus occur. In the temperature range 1100–1300°C, copper forms a molten phase as soon as its limited solubility in austenite is reached.[23] The level of surface concentration of copper and the concentration gradient in the bulk steel depend on time, temperature and furnace atmosphere. Copper concentration at the surface can easily exceed its solubility in iron, which at 1100°C is only about 9% so that at temperatures exceeding the melting point of copper (1083°C) the metal appears as a layer of pure copper between the steel and the scale. Any stretching of the surface layers then provides cracks at least as long as the depth of molten copper. Further attack at grain boundaries freshly exposed by the stretching can create yet deeper fissures.

If the hot working temperature is in the range 1040–1150°C, small changes in the melting point of copper will determine whether or not it is molten, and will therefore influence surface cracking. To the extent that nickel and molybdenum raise the melting point of copper, they retard cracking. Conversely, tin and antimony, which lower the melting point, increase the sensitivity of steel to surface cracking; their influence seems to act entirely through copper, the key element, in whose absence they have little adverse effect. Arsenic (slightly deleterious) and nickel (slightly beneficial) are of such small importance as to be negligible.

The effect of tin in reducing copper solubility in austenite may be the overriding mechanism. It is reported that 2·4% of tin in iron at 1100°K can reduce copper solubility to 3%[24] and that 1·0% of Sn can reduce it, at 1200°K, by a factor of two.[25] A molten Cu-rich phase will then appear at a much lower level of enrichment than would be the case in the absence of tin. The effects of arsenic and antimony may be similar to that of tin.

It has been suggested that diffusion into the bulk steel at high soaking temperatures might outstrip the rate of copper enrichment, with a critical dispersion temperature varying from 1413°K at 0·05% to 1573°K at 0·2% Cu. However, such a procedure is seen to place one further constraint on the thermal treatment to be applied to the steel, and is an adequate reason for attempting to minimise the copper content.

Other work indicates that, in the presence of sulphur, nickel additions can counteract the effects of copper; steels containing 0·35% Cu and 0·33% Ni were virtually crack-free, owing to the action of nickel in raising the melting point of copper and of increasing the solubility of copper in austenite.[26]

[23] A. Nicholson and J. D. Murray, Surface hot-shortness in low-carbon steels, *J. Iron Steel Inst.*, **203**, (10), Oct. 1965.
[24] B. B. Hundy, *op. cit.*
[25] D. A. Melford, Surface hot shortness in mild steel, *loc. cit.*
[26] *A. Nicholson and J. D. Murray, loc. cit.*

It has also been noted that the presence of silicon in large enough concentrations can promote the formation of iron-rich rather than copper-rich alloys on the surface.[27]

Copper arises in ferrous scrap largely from residual wiring. In the future, it may arise from the use of scrap of those steels to which copper has been added to confer corrosion-resistance.

Copper removal from steel scrap continues to be of interest. A process under development in the U.K. involves use of a reagent which absorbs significant quantities of copper; details have not yet been disclosed.[28] A Czechoslovakian patent in this field relates, however, not to the upgrading of the steel but to the recovery of an alloy containing 98–99% Cu, in a sweating process not dissimilar to that described many years ago by Leak and Fine.[29, 30]

4.4.3.2 *Tin*

Like copper, tin is not removed from steel during normal refining operations.

It is evident that the most harmful effects of tin on the impact properties of steel are apparent only in the presence of copper, which is normally present in mild steel to a concentration 4–8 times that of tin. A limit of 0·6% SN has been given for satisfactory cold working of mild steel under laboratory conditions; at greater tin contents, intergranular cracking is likely.[31] At 0.4% Sn, impairment of the impact strength and raising of the ductile-brittle transformation temperature are observed under certain conditions; some of the effects are ameliorated by nickel additions. It is concluded that tin, at least up to the 1% level, behaves as a normal solid-solution hardener in annealed mild steel, but its synergistic effects with copper must be remembered.

In fact, because of its deleterious effects on impact properties, few specifications permit 0·08% tin, and most are of the order 0·02–0·04%; iron ores are generally low in tin and contamination arises principally from scrap, especially from incinerated tin-plate. In auto scrap, tin can arise from bearings.

Tin may be removed selectively from tin-containing residues, such as copper wire, by treatment with molten alkali metal hydroxide, preferably

[27] Mark I. Copeland and John S. Howe, *Preventing formation of copper alloys of tin, antimony and arsenic on steel surfaces during reheating to reduce hot shortness*, Washington, D.C., US Bureau of Mines, RI 8080, 1975.

[28] *BNF News*, No.44, March 1988. (BNF Metals Technology Centre, Wantage.)

[29] Vance G. Leak, Morris M. Fine and Henry Dolezal, *Separating copper from scrap by preferential melting*, Washington, D.C., United States Department of the Interior, Bureau of Mines, RI 7809, 1973.

[30] Vance G. Leak and Morris M. Fine, Recovery of copper and steel from scrap, U.S.Pat. 3776 718, Filed Jul. 13 1972.

[31] A. B. Shelmerdine and D. A. Robins, Influence of tin on the mechanical properties of iron and steel, *J. Iron Steel Inst.*, **203**, (1), Jan. 1965.

NaOH or KOH. The advantages of this method over the well-known conventional de-tinning methods for tinplate and employing aqueous solutions of NaOH are not stated.[32]

For several reasons, not excluding a (temporary) high price of tin, much attention has been turned to its substitution by other materials.[33] A notable example is the use of Tin-Free Steel (TFS) in cans. Hence, it is possible that the problem of tin may be less acute in the future.

4.4.3.3 *Aluminium*
Aluminium is readily oxidised and enters the slag, with a possible danger of increased refractory wear and, in the blast furnace, of phosphorus reversion. In the metal it is harmless, and is commonly added as a deoxidant; it may, in that capacity, increase the inclusion count.

4.4.3.4 *Lead*
Small amounts of lead in steel improve its machinability, and additions are made for that purpose. It may also arise in auto scrap from fragments of battery plates or body-filling metal. Though use of the latter is fast declining, some use persists.[34]

Lead has low oxide stability but is almost insoluble in iron and its alloys. It therefore poses no problems in the electric arc furnace or basic oxygen process, where some 97% of the lead charges is recovered in the dust; subsequent use of this dust for blast furnace sinter could, however, affect blast furnace refractory life.[35] Accumulations of lead below the salamander[36] in the blast furnace pose a particular risk of a break-out through the refractory.

4.4.3.5 *Nickel*
Some effects of nickel have already been noted in connection with sulphurous atmospheres and with amelioration of the effects of copper and tin. Generally, nickel is added for the increased hardenability that it confers. Its incidence is likely to increase with the increased use and recycling of those alloy steels that are not readily segregated from plain carbon grades. In particular, the automotive use of HSLA (high-strength, low-alloy) steels is increasing.

4.4.3.6 *Phosphorus*
Phosphorus, like sulphur, is almost always present in blast furnace iron and is similarly undesirable, although it improves the fluidity of cast irons

[32] Shuichi Odo and Tetsuo Yamaki, Selective recovery of tin from coatings on scrap alloy, *Jpn. Kokai Tokkyo Koho* JP 62 93,319 [87 93,319] (Cl.C22B25/04) Apr. 28 1987.
[33] John E. Tilton (Ed.), Materials Substitution: Lessons from the Tin-Using Industries, Washington, D.C., Resources for the Future, Inc., 1983.
[34] Patrick D. Canavan, Solder, Chapter in Material Substitution: Lessons from Tin-Using Industries, Washington, D.C., Resources for the Future, Inc., pp.36 *et seqq*, 1983.
[35] E.J. Ostrowski, Recycling of tin free steel cans, tin cans and scrap from municipal incinerator residue, *Proc. Amer. Iron Steel Inst.*, 79th General Meeting, New York, May 26, 1971.
[36] Otherwise known as a 'bear' or 'horse', this is a mass of metal found below the hearth level of a blown-out blast furnace, resulting from penetration of the refractory by molten metal.

and the corrosion resistance of some steels. Free-machining steels may contain up to 0·5% S and 0·15% P.

Phosphorus is readily removable in standard basic oxygen and electric arc steelmaking operations. However, a new development is the use of re-phosphorised steels whose widespread introduction may well pose questions regarding the availability of sufficient basic steelmaking capacity.

4.4.3.7 *Sulphur*

The presence of sulphur, with consequent sulphide formation at grain boundaries, increases the likelihood of hot tearing and hot cracking. It also lowers the room temperature ductility of many steels. A beneficial effect is its improvement, by formation of manganese sulphide, of machining characteristics, and this might improve the overall economics of product manufacture in cases where some loss of ductility can be tolerated. Sulphur removal is facilitated by very basic slags, particularly under the reducing conditions attainable in the basic electric furnace.

4.4.3.8 *Zinc*

Zinc arises from galvanised material and from brass. It is largely oxidised in steelmaking, and the voluminous dust so produced is a major problem in the gas regenerators of the open-hearth furnace and in the gas-cleaning system of the electric arc furnace. The physical properties of these dusts make them difficult to handle, but they are potential sources of lead and zinc; these metals preclude their return to the sinter plant since, although most of the zinc subsequently entering the blast furnace reports to the dust and slag, sufficient can accumulate as accretions of oxide to constitute a serious risk of scaffolding[37] and of accelerated refractory wear.[38]

4.4.3.9 *Other metallic contaminants*

Residual chromium, manganese and molybdenum originating in alloy steel scrap can measurably increase the hardenability of medium carbon steels. Since this effect is usually desirable this group of alloying elements is added for that purpose to certain grades of steel. Indeed, alloy steel scrap that contains them is purchased as an inexpensive source of alloying additions. However, other elements, such as arsenic and antimony, are extremely deleterious to, for example, the performance of those steels formulated for enhanced resistance to temper-embrittlement. It is especially important that steels for use in power station forgings, at service temperatures of 450–550°C, should be made from ores that are low in these elements. Since such ores are in relatively short supply there is a good case for segregating rotor forgings from other power station scrap and for reusing them in similar applications.

---

[37] Sticking of the burden to the side of the blast furnace, so reducing the effective area of the stack.
[38] N.G. West, Recycling ferruginous wastes, *Iron and Steel International*, June 1976.

At the levels likely to be encountered with prevailing scrap-utilisation rates none of the other metallic tramp elements reporting principally to the steel has been identified as the source of serious complications during refining, fabrication or service.

Account must be taken of newer contaminants which may be entering the materials supply stream. Metals such as gallium, samarium and yttrium can conceivably arise in computer and other semiconductor scrap, albeit in small concentration. Yttrium also has some application in automotive exhaust systems. It is possible that some of these, singly or in combination, may have unforeseen effects on the properties and micro-structure of the steels into which they may pass.

4.4.4 Permissible limits for residual elements in steel

It is clear that no consistent limits may be laid down for residual elements in steels. Any residual may improve some properties, impair others and may operate synergistically with other residuals. The level of impurities to be tolerated in a steel must therefore be determined by consideration of the purpose to which the steel is to be put.

Integrated plants generally make steel for a wide variety of products and applications. It is not practicable to assign specific melting facilities to the specification production of iron for an individual product in order to control minor elements that might not be removed in steelmaking. Thus, the choice of an iron ore from the very wide range of compositions that is, in principle, available, or the adjustment of steelmaking practice to provide material for a few selected applications, would result in a severe loss of flexibility.

Maximum flexibility requires that the raw materials used must contain no tramp elements that will have an adverse effect on the manufacture or properties of a major product of the plant. In practice, the production of a wide range of products in a given plant requires raw materials that will meet the most stringent demands of any of its major products. A specifi-cation for one major product of $0 \cdot 1\%$ Cu maximum implies the organisa-tion of the raw materials supply to meet this specification throughout, even though other plant products might be able to tolerate considerably higher levels.

The whole subject of residual levels is a matter of some delicacy to steelmakers, and data are sparse. The care with which manufacturers guard their internal specifications has been attributed to a number of commercial factors relating to their competitive position. A degree of defensiveness is also apparent. Published data should therefore be inter-preted with caution. In order to obtain a greater margin of safety, a purchaser may often specify a lower level of a particular element than is common practice for the application in question.

Examination of a very large number of specifications of residual max-ima of those elements that are likely to arise from scrap suggests that the

Table 4.2
Typical residual element levels in steel

Application	Ni	Cr	Cu	Sn	Additional limitations	Percentage of total tonnage
Extra deep drawing	0·10	0·03	0·10	0·03	0·18 total	Negligible
Fine wire	0·10	0·06	0·12	0·04	0·20 total	11
Coiled plate (excluding extra deep drawing)	0·20	0·10	0·20	0·05	—	11
Beams, sections	0·20	0·15	0·30	0·05	Cu+Sn 0·30 max.	61
Reinforcing bar	0·30	0·30	0·30	0·08	None	17

number specifying <0·10% Cu is actually very small indeed, as is also tje number requiring <0·03% Sn. Typical specifications for one U.K. works are given below:

The copper and tin are controlled specifically to limit surface hot shortness.

It should be noted that some 78% of the output cited in Table 4.2 can tolerate 0·20% Ni, 0·15% Cr, 0·30% Cu and 0·05% Sn. Even deep drawing grades tolerate, typically, 0·20% Cu, 0·20% Ni and 0·03% Sn.

Permissible sulphur and phosphorus levels in carbon and carbon-manganese steels are generally 0·05% maximum for each element but are governed by intended use and no rigid rule exists. Limits are imposed, also, by the American Iron and Steel Institute on the residuals in steels produced by different processes, as follows:[39]

Basic electric 0·025% each S and P
Basic open hearth or oxygen 0·04% S, 0·035% P
Acid open hearth or electric 0·05% each S and P

It is evident that the permitted maxima cover a wide range. It also appears that some 78% of the output cited in Table 4.2 can tolerate 0·20% Ni, 0·15% Cr, 0·30% Cu and 0·05% Sn. Even deep drawing grades tolerate, typically, 0·20% Cu, 0·20% Ni and 0·03% Sn. Tube-making grades commonly specify <0·015% Sn and <0·12% Cu, particularly in continuously welded types, where the tube is hot-formed and where significant concentrations of residuals at the interface could impair welding.

[39] *Iron and Steel Specifications*, London, British Steel Corporation, 4th edition, 1974.

Generally, the blastfurnace iron produced in the U.S.A. or U.K. is not the limiting factor in achieving residual levels of this order. The principal source of tramp elements is the scrap, where the aim is to maintain a copper content of 0·15% maximum in the scrap, so facilitating steelmaking flexibility.

It is clear, therefore, that low-grade scraps have only limited acceptability in some markets. The potential of various market sectors for absorbing low-grade scrap is given in Table 12.1.

4.5 CONTAMINATION LEVELS IN STEEL SCRAP

It is axiomatic that the steelmaker seeks scrap of minimum contamination. In order to produce steel of a given residuals content it is necessary to charge hot metal, iron ore and scrap of known purity.

Scrap is a recirculating load; each time that it is recycled it can introduce fresh contamination, cumulatively raising the impurity level of the basic steel scrap. Such stock, when eventually it forms scrap, is of ever-diminishing acceptability.

4.5.1 Baled scrap

Obsolete scrap, including car hulks, compressed into a bale, carries with it all the non-ferrous and non-metallic contamination that the dismantler and/or processor had not found it worthwhile or convenient to remove. As one of the lowest grades of scrap it has been the worst affected by the reduction in demand for low-quality material by the steel industry. This in turn has had a significant backward link to the discarded products, such as road vehicles, which are a major source of obsolete scrap.

The definition of the No. 5 bale (No. 2 bundle in the U.S.A.) simply stipulates old, black (uncoated) and galvanized material, hydraulically compressed to charging box size, and weighing not less than 1200 kg/m³.

The composition is extremely variable. To the steelmaker, it is simply a compressed bale of unknown origin and of a composition and yield that cannot be determined short of melting and analysis. It may comprise an entire vehicle with all components, including tyres, battery, copper wiring etc., or it may be a stripped body hulk or any other light-gauge steel scrap, such as water heaters, refrigerators and other appliances.

4.5.1.1 *Economic factors*

The importance of scrap quality: has been stressed, and the processing method determines the level and variation in bale contamination. Attainment of high quality in baled scrap entails extensive hand-stripping of potential contaminants. This is feasible only if the added cost can be recovered from the additional revenue generated by higher demand or higher price for the product, or from the sale of the non-ferrous contaminants. It has usually been considered that the added cost would be greater than the added value, i.e. in the opinion of the consumers insufficient value is added to the stripped product to interest them in paying the premium price that would justify the additional effort by the scrap processor. The problem arises because of the way in which contaminants are disseminated throughout the material that comprises obsolete scrap.

The reason why most steel companies are unwilling to pay a premium seems to lie in the high risk factor associated with the difficulties of adequate analysis of the improved bale. However, certain U.S. scrap dealers have negotiated contracts with steelmakers who pay a premium for high-quality bales. This normally requires that a staff member from the steel company should work with the processor to ensure a standard stripping procedure. The company then purchased all the output from the processor, paying the price for No. 2 Heavy Melting Scrap, instead of that for the No. 2 bundle. Certain U.K. scrapyards also operate such arrangements.

4.5.1.2 *Physical factors*
The size of the bale precludes its use in the cupola;, as employed in iron foundries. The average diagonal of a bale is around 1·35 m; hence a cupola with a throat diameter of around 4 m would be needed to melt such material, implying a furnace producing around 90 tonnes/h, compared with typical foundry cupolas producing about 45 tonnes/h.

4.5.1.3 *Chemical factors*
The possible contamination level of the bale is evident to any visitor to a yard where baling is performed. Scrap is lifted by crane from the stockpile and is dropped into the charging box; thus, contamination inherent in the scrap itself is joined by that arising from contact with the ground and with oil and petrol which spill from hulks.

Bales are noticeably heterogeneous; visual inspection of the surface shows that they may contain large pieces of wood, whole car tyres, building dbris, and many other types of contaminant.

The only available statistics on the composition of the bale are those determined almost thirty years ago by Battelle Memorial Institute.[40] The results are given in Table 4.3. They show that the average content of both copper and tin is unacceptably high for remelting for most purposes. Of even greater significance, the standard deviation was high for each element.

From published data we may draw important conclusions about the composition of baled scrap:

(a) The density of the bale can be much less than the stipulated minimum of 1200 kg/m^3.

---

[40] Battelle Memorial Institute, *The measurement and improvement of scrap quality,* Report to the Institute of Scrap Iron and Steel, Inc., Columbus, Battelle Memorial Institute, Dec. 23 1960.

Table 4.3
Copper and tin content of baled scrap (Battelle)

Element	Mean	Standard deviation	Maximum
	%	%	%
Copper	0.48	0.44	2.79
Tin	0.08	0.07	0.285

(b) Yield losses from the inclusion of wood, plastic, fabric, glass etc. can range from 3·9 to 38%.

(c) The probability that a given melt will exceed a stipulated limit in any residual depends on the average level in the bale and also on its standard deviation. In any hypothetical melt of known quantities of hot metal, scrap of known purity, and No. 5 bales (No. 2 bundles), a reduction of the standard deviation of the copper content of the bale from 0·44% to 0·15%, while maintaining an average copper content of 0·48%, may be calculated to reduce the chances of exceeding an arbitrary residual copper content from 20 to 10%.

(d) The steelmaker will adjust use of iron sources according to the lowest level of quality that experience has taught him to expect, and great practical significance then attaches to the standard deviation, which is indicative of the proportion of off-specification heats that will be produced. Some of these heats will necessarily be discarded and the direct financial loss will be the greater if alloying additions have already been made.

(e) The savings attainable through the use of No. 2 scrap instead of No. 1 are small compared with the potential loss to the steelmaker, should an entire cast be ruined.

4.5.2 Shredded scrap

The scrap shredder, otherwise known as the fragmentiser or pulveriser, was introduced into the secondary metals industry in 1961, and its growth has been rapid.

Size reduction is effected by a combination of impact, tensile and shear stresses; the product comprises fragments in the size range 2–30 cm.

4.5.2.1 *Financial factors*

Shredded scrap has several important financial and techno-economic advantages over bales. These arise from its physical and chemical characteristics.

4.5.2.2 *Physical factors*

(a) Shredded scrap is much smaller in size than is baled. Hence, it is easier to inspect for non-ferrous contamination.

(b) A degree of magnetic decontamination is possible.

(c) The size range is acceptable to cupolas as well as to steelmaking furnaces.

(d) The high surface area/volume ratio facilitates heat transfer and ensures rapid melting.

(e) The physical form of shredded scrap greatly simplifies handling procedures.

The feed to the scrap shredder can be very diverse, including cars which are complete except for the petrol tank; however, to minimise hammer wear, the lighter gauges of scrap are preferable.

4.5.2.3 *Chemical factors*

Data on the composition of shredded scrap are available from several sources; none, however, provides the statistical information necessary to assess the probable limits of composition. Furthermore, although some sources state that only automotive scrap was fed to the shredder, many fail to characterise the input. It is therefore difficult to assess the degree to which the data represent the total output of shredded scrap.

From 1964 onwards, General Motors investigated the role of cars in scrap utilisation, presenting data on shredded scrap from such well-known concerns as Luria Brothers and Proler.[41] It also examined the output of another shredder operator, Sam Allen and Sons, at Pontiac, Michigan; at this company, the feed was first treated in a slow-speed mill, termed a pre-shredder. The mill, driven by a motor of only 200 horse-power compared with the 1750–10 000 horsepower of the main shredder, tore the car hulk into 8–10 large pieces which were then reduced further in size in an hammermill-type shredder. The prime function of the pre-shredder was the elimination of power surges in the main shredder, reducing the costs of energy and of maintenance. However, although not specifically so intended, by breaking up the hulk into loose pieces amenable to further shredding it minimised the entrapment of wire and other contaminants that are often encapsulated in shredded scrap.[42]

From comparison of published data it is immediately obvious that shredded scrap has a substantially lower copper 0·20–0·22%, compared with 0·48–0·55% for baled scrap. Of possibly greater importance is its lower standard deviation, which is only about 0·04%, compared with the 0·44% for bales. The material can thus be used with a greater degree of confidence that the resulting melt will be within specification. It is also lower in tin.

The reason for the superior performance of the shredder in producing high-purity scrap may be found in its mode of operation. Whereas the baler compresses its entire charge into a dense form which defies further efforts to remove contaminants, the shredder performs essentially the

[41] F. J. Uhlig, General Motors' efforts in junk car processing, *Proc. 52nd National Open Hearth and Basic Oxygen Steel Conf., Am. Inst. Min., Met., and Petroleum Eng.*, Open hearth proceedings, pp.83–90, 1969.

[42] Ripsteel Corporation, private communications.

same function as does the ore crusher in liberating valuables from gangue. The shredded scrap is amenable to magnetic separation with an improved chance that the ferrous and non-ferrous materials will be present as discrete particles.

The available data on shredded scrap cannot be termed of good statistical quality since they are generated under conditions that are often unspecified. Undocumented data by Battelle,[43] Burlingame[44] and Sawyer[45] are sufficiently similar to suggest a common origin. Confidence limits are not given.

However, shredded scrap has rapidly gained in acceptance and performs well as a portion of the charge for producing both steel and cast iron. It is not, though, always felt by foundrymen that the added value of shredded steel is commensurate with its additional cost. In particular, it is considered that for a price which approaches that of some No. 1 grades it has unreasonably high levels of chromium, molybdenum, and nickel which generally originate in alloy steels, stainless irons and plated or clad steels, all of which report to the magnetic fraction.

The further improvement of shredded scrap is difficult, since in order to remove tramp element levels to the equivalent of those found in quality cold-forming steel sheet, some of the shredded scrap must be separated out as alloyed fractions. No existing technology can effect such a separation under current economic conditions. Further, there appears to be little incentive to seek copper levels lower than the 0·15–0·25% level attained by shredding, since existing shredding and cleaning processes seem to assure shredded scrap, even at 50% premium, of a steady market relative to the No. 5 bale/No. 2 bundle.

4.6 THE STEEL SCRAP INDUSTRY

4.6.1 Introduction

The greatest long-term problem of the scrap industry is that of cumulative contamination.

The views of one very large scrap dealer are of relevance: so far as automotive scrap is concerned, it is suggested that around 99% of all cars scrapped in the U.K. now go through shredders. Though not everyone would agree with this opinion, it is clear that the shredder processes a large fraction of scrapped vehicles. Some of the scrap is then hand-sorted to remove non-ferrous contaminants for a particular long-term contract which guarantees a premium price for properly stripped scrap, i.e. sub-

[43] W. J. Regan, R. W. James and T. J. McLeer, [Final report on] *Identification of opportunities for increased recycling of ferrous solid waste*, Report by Battelle Memorial Inst. to the Scrap Metal Research and Education Foundation of the Inst. of Scrap Iron and Steel, Inc., Washington, D.C., United States Environmental Protection Agency, (SW-45d)(PB 213 577), 1972.

[44] Richard D. Burlingame, *loc. cit.*

[45] James W. Sawyer, Jr., *Automotive scrap recycling: processes, prices and prospects*, Baltimore and London, The Johns Hopkins University Press for Resources for the Future, Inc., 1974.

stantially free from dirt and non-ferrous metals. The mixed No. 2 grade material can, by hand-sorting, achieve a composition approximately equal to that of fragmentised scrap.

Certain private-sector U.K. steelmakers pay a premium for shredded scrap of enhanced quality. The material sent to one such consumer is claimed to be totally free from dirt, non-metallics and non-ferrous metals. The company concerned consumes 14 000 tonnes of scrap steel per week and each week carries out melting trials on charges comprising 50% its own captive scrap and 50% material from its contracting scrap merchant. The steelworks has formerly encountered high contamination levels in all charges, and fragmentised scrap is too expensive for the relatively low-grade application in question. The premium actually paid over the standard price for No. 2 is said to be close to the additional cost of the hand-sorting, i.e. labour costs plus losses, largely because of the tonnage loss in hand-sorting and because the material purchased is more intensively processed, e.g. it is magnetically separated three times. The gain to the merchant is in continuity, in that the material is saleable at all times, whereas the demand for No. 2 is intermittent. This is an individual arrangement, still [1986] nominally on a trial basis of 300 tonne/day.

4.6.2 Supply and demand for different grades of steel scrap

It has been noted that the basic oxygen process, which provides a large proportion of the steelmaking capacity of the industrialised countries, can be supplied with scrap almost entirely from steelworks' arisings. This picture will, however, be distorted as continuous casting makes further inroads into traditional casting methods.

The acceptability of purchased scrap is, partially at least, limited by the rapid nature of the basic oxygen steelmaking process, which leaves little margin for error. Any increase in the percentage of low-grade material increases the risk of exceeding residual specifications. Once the point of excessive contamination is reached, no price reduction for old scrap will induce producers to use it instead of higher quality scrap. Since it is so poor a substitute for other grades, a reduction in its real price will cause the demand to remain constant or to increase relatively little, i.e. the demand function is price-inelastic. On the other hand, a rise in the real price may produce large reductions in demand as other, higher quality scraps are economically substituted for baled obsolete scrap.

The electric arc furnace uses 98% scrap, and falling scrap prices were favourable enough to cause its contribution to total U.S. production to increase, from 2% to 20% in the period from 1960 to 1985.[46] Recent advantages in electric steelmaking have been concentrated in ministeel mills, turning out 50 000–500 000 tonnes/year. Their staple product is reinforcing bar. At suitably low scrap prices the regional electric furnace shop can evidently compete effectively within a limited radius. Its burden

[46] Barnett and Crandall, *op. cit.*, p.12.

is almost wholly scrap, which need not be of high quality, thus offsetting somewhat the reduction in the use of low-grade scrap by traditional steel mills.

Many mills still purchase bales provided that they know them to have been well-prepared, which implies an established relationship with the supplier, and provided that their price is competitive. It must, however, be concluded that bundled scrap is in regular demand only at times of peak steelmaking activity.

4.6.3 Alternative markets for low-grade scrap

The mingling of metals, by alloying and complex construction, continues inexorably. Data for tin and copper contamination at one plant demonstrate a marked rise in their residual levels since World War I.[47] There are, however, applications that will tolerate higher residual levels than those discussed earlier. Reinforcing bar provides one such case. The final ferrous metallic sink is cast iron.

[47] Hundy, *op. cit.*

5
THE RECOVERY OF MATERIALS FROM MOTOR VEHICLES

5.1 INTRODUCTION

Many studies have been made of the social, financial and environmental aspects of dealing with obsolete cars.[1-5]

In the developed countries, discarded road vehicles form a major source of the post-consumer ferrous scrap whose availability, unlike that of grades generated by steelmakers or users, is significantly affected by price. They are also major sources of secondary aluminium, copper, lead and zinc. However, they also generate another fraction, which is not generally valuable, namely the non-metallic *detritus* arising from seats, carpets, tyres and other components.

Unfortunately, recovery of the valuable materials in the motor car is not simple. The vehicle is complex and self-contaminating, characteristics which can adversely affect recoverable values. For example, failure to separate the non-ferrous metals from the steel means that the latter must be classified as low-grade material, and can reduce the value from a total of more than $108/car, as separate fractions, to $65 as a single, contaminated grade.

Moreover, efficient separation of the materials of the vehicle cannot be performed without cost. So far as the individual scrap processor is concerned the pertinent question is whether the sum of all the costs exceeds the sum of all the revenues. That question is also of concern to the scrap consumer. Both processor and consumer are normally concerned only with *individual* costs.

Motor vehicles entering the recycling system will include buses, lorries and passenger cars; material from these categories is generically termed 'automotive scrap'. A scrap vehicle arises when it no longer satisfies the

[1] Robert Louis Adams, *An economic analysis of the junk automobile problem,* Washington, D.C., U.S. Department of the Interior, Bureau of Mines, IC 8596, 1973.
[2] Staff, United States Bureau of Mines, *Automobile disposal, a national problem,* Washington, D.C., U.S. Department of the Interior, Bureau of Mines, 1967.
[3] Management Technology Inc, *Automobile scrapping processes and needs for Maryland,* Washington, D.C., United States Department of Health, Education and Welfare, Bureau of Solid Waste Management, 1970.
[4] Anon, *The Automobile Cycle: an environmental and resource reclamation problem,* Washington, D.C., United States Environmental Protection Agency, (SW-80 ts.1), 1972.
[5] National Industrial Pollution Council, Sub-Council Report: *Junk Car Disposal,* prepared for the Secretary of Commerce, Washington, D.C., United States Government Printing Office, 1971.

needs of its owner for transport. This may be because it is no longer feasible to keep it in running condition, as defined by the appropriate licensing authorities. The enforcement of emission requirements may result in earlier scrapping of vehicles. The proper description of the vehicle at that point, whether abandoned,[6] deregistered, derelict,[7] discarded, junked, obsolete, or wrecked, has been the subject of much debate that lies outside the scope of this study. It is proposed, hereafter, to use the term 'discarded' to denote disposal by the owner.[8]

The so-called 'scrap car problem' refers to the stocks of unprocessed, discarded vehicles. Before the problem arose, the last owner of the vehicle would either abandon it or deliver it to the recycling system by taking it to a dismantler who, after removing profitable parts or metal, would deliver the stripped carcase [hulk] to a scrap processor. It would then usually be compressed into a cuboid or rectangular bale known in the U.K. as a No. 5 bale and in the U.S.A. as a No. 2 bundle.[9] Sometimes the hulk was burned to remove the non-metallic items but, since this step involved cost, it was often omitted.[10] The bale would then go to a domestic or foreign steel mill to be charged, with pig iron and other types of steel scrap, to produce new steel; it carried with it all the non-ferrous and non-metallic contaminants which it had not been worthwhile to remove. Alternatively, the hulk was partially compressed into a so-called scrap log, which was then sheared into slabs of size convenient for handling and for charging into the furnace. Some contaminants were usually liberated during shearing. As noted in Section 4.5.1, a baled vehicle is not a desirable source of iron units except in periods of scrap shortage.

Early in the 1960s the market was considerably disturbed by the advent of the scrap shredder. For reasons already discussed (4.5.2), the purity of shredded scrap makes it acceptable to steel mills whereas that of the bale is often in question. Furthermore, its small particle size renders it suitable as feed for the cupola furnaces of the iron foundry, for which the bale had been too large. In the late 1960s, the cost of transporting hulks was reduced by the portable flattener which, by reducing the size of the hulk, permitted many more of them to be accommodated in a single load.

Meanwhile, the steel industry was itself undergoing changes that were to have radical effects on the markets for scrap (see Section 4.3).

---

[6] Abandoned vehicle: a vehicle left unattended for more than 48h on private or public property, without notification to the land owner or appropriate law-enforcement agency.

[7] Derelict vehicle: one that has not been reregistered (re-licenced), with the intention of keeping it permanently deregistered, and that is in a wrecked, worn-out, extensively-damaged, dismantled or inoperative condition.

[8] The importance of precise terminology lies in the legalities of the removal and processing of apparently abandoned, ownerless vehicles which might possibly have been stolen.

[9] In US parlance No. 1 denotes high quality scrap and No. 2 a scrap of lower purity or of a shape leading to high melting losses.

[10] The Clean Air Acts, and similar legislation in other countries, have in principle made the burning of hulks, as of copper cable, impossible except in incinerators of a type possessed by only a few scrap dealers. The law has, on occasion, been circumvented by operating after dark.

In the final analysis, the reason why vehicles are or are not processed to secondary metals is an economic one. Since 1968, a number of major studies have been made of the economic and technological problems of the recycling of cars. The principal geographical origin of the vehicles is given:

(a) Stone (U.S.A., 1968): an analysis of the finances and technology of further stripping of a hulk from which 75–90% of the copper had already been removed by dismantlers; it was assumed that no further non-ferrous revenue was obtained, and the economic case was stated entirely in terms of improving the quality of the steel.[11]

(b) Dean and Sterner (U.S.A., 1969): this study presented data for the complete stripping of cars, in order to examine the profitability of a dismantling operation deriving revenue from the sale of No. 2 bundle steel, cast iron and non-ferrous metals.[12]

(c) Adams (U.S.A., 1973): an econometric study which, using the dismantling data of Dean and Sterner (*q.v.*), assessed the profitability of baling relative to that of shredding, having regard to the degree of completeness of the car on arrival at the scrapyard.[13]

(d) Sawyer (U.S.A., 1974): an economic examination of the supply of car hulks and, via a computer-aided dismantling model, of the feasibility of reducing residuals to low levels. It, too, was based on data of Dean and Sterner.[14]

(e) Roig et al. (U.S.A. and imported, 1975): a phenomenological and financial analysis of the effects of design changes on the recovery of materials from scrapped road vehicles.[15]

(f) The American Society of Mechanical Engineers (U.S.A., Europe & Japan, 1984), provided a compilation of data on the materials presently and potentially used in road vehicles.[16]

Other, more specialised, sources are:

(g) Sterner *et al.* (Japan, 1984) extended the classic work of Dean and Sterner to four makes of Japanese cars commonly imported into the United States.[17]

[11] Ralph Stone and Company, Inc., *A survey and analysis of the supply and availability of obsolete iron and steel scrap,* Report to the Business and Defense Services Administration, Washington, D.C., U.S. Department of Commerce, 1957.

[12] Karl C. Dean and Joseph W. Sterner, *Dismantling a typical junk automobile to produce quality scrap,* Washington, D.C., United States Bureau of Mines, RI 7350, 1969.

[13] Robert Louis Adams, *op. cit.*

[14] James W. Sawyer, Jr., *Automotive scrap recycling: Processes, prices and prospects, Baltimore and London, The Johns Hopkins University Press for Resources of the Future, Inc., 1974.*

[15] Robert W. Roig, Marc Narkus-Kramer and Andrea L. Watson, *Impacts of materials substitution in automobile manufacture on resource recovery, Symp., The technology of automobile recycling,* University of Wisconsin, Oct. 16 1975.

[16] Various authors, *The Impacts of Material Substitution on the Recyclability of Automobiles,* New York, The American Society of Mechanical Engineers, 1984.

[17] J.W. Sterner, D.K. Steele and M.B. Shirts, *Hand dismantling and shredding of Japanese automobiles to determine material contents and metal recoveries,* Washington, D.C., U.S. Department of the Interior, Bureau of Mines, RI 8855, 1984.

(h) Martini (Italy, undated) related entirely to certain models of FIAT.[18]

(i) Henstock (Europe, 1982), on the basis of eleven published or communicated analyses of the composition of European cars, examined the implications for the secondary aluminium industry of an increased availability of scrap from vehicles.[19]

(j) Niimi (Japan, 1984) presented the statistics of automotive materials consumption, and of vehicle scrapping in Japan.[20]

A summary of United States Bureau of Mines work on vehicle recycling is now available.[21]

5.2 RECOVERABLE MATERIALS IN THE MOTOR VEHICLE

Many of the available data on the composition of the automobile originate in the U.S.A. Hence, the financial data are given here in U.S. dollars.

In 1981 the motor vehicle industry of the U.S.A. absorbed the following percentages of U.S. consumption of the stated materials:[22]

Table 5.1
Materials consumption by the motor vehicle industries of the U.S.A. (1981)

Material	(%)
Total steel	18·6
Malleable iron	39·2
Aluminium	15·8
Copper and alloys	11·0
Cotton	0·7
Plastic resin	3·9
Natural rubber	78·2
Synthetic rubber	56·6
Zinc	27·7
Lead	45·2

[18] P. Martini, private communication, 1986. The undated data relate to certain models of FIAT.

[19] Michael E. Henstock, The European Picture, chapter in *The Impacts of Material Substitution on the Recyclability of Automobiles,* New York, The American Society of Mechanical Engineers, pp.171–185, 1984.

[20] Itaru Niimi, The Japanese Picture, chapter in *The Impacts of Material Substitution on the Recyclability of Automobiles,* New York, The American Society of Mechanical Engineers, pp.187–198, 1984.

[21] K.C. Dean, J.W. Sterner, M.B. Shirts and L.J. Froisland, *Bureau of Mines Research on Recycling Scrapped Automobiles,* Washington, D.C., U.S.Department of the Interior, Bureau of Mines, Bulletin 684, 1985.

[22] *MVMA Motor Vehicle Facts & Figures,* Detroit, Motor Vehicle Manufacturers Association of the United States, Inc., 1981.

Thus, discarded vehicles are potential sources of significant quantities of secondary materials.

In a 1969 paper that has become a standard reference, Dean and Sterner examined the components of fifteen cars manufactured between 1954 and 1965 and determined their composition in terms of aluminium, cast iron, copper, glass, lead, rubber, steel, zinc, and miscellaneous. Their aim was to determine whether the non-metallic portions of a typical vehicle could economically be burned out in a smokeless incinerator and the residue hand-dismantled to yield steel that would form a high-quality No. 2 bundle.

The work showed that a hypothetical car, whose composition was based on a statistical, sales-weighted analysis of fifteen vehicles then representative of those arriving at U.S. scrapyards, could be processed into $56 worth of marketable ferrous and non-ferrous metals. At the prevailing wage rates, its processing would have cost $51, and the hypothetical scrapyard could have yielded an annual rate of return on investment of 19%.[23]

Care was taken to ensure that the vehicles eventually selected for analysis were representative of a 'composite' vehicle, whose composition was computed from that of the individual vehicles. Factors considered were:

(a) statistical data gathered by the U.S. Department of Commerce on the ages of obsolete cars,

(b) makes and body types according to the numbers originally manufactured, and

(c) auxiliary equipment, such as air-conditioning, power brakes etc.

Examples of the fifteen cars finally selected were purchased from dismantlers and time and motion studies were made of the times required to:

(a) remove components from the car, and
(b) disassemble them into saleable metallic fractions.

The results were supplemented by time and motion data from a local scrapyard engaged in the dismantling of incinerated cars into No. 2 grade steel, cast iron, and non-ferrous metals. The main components were then analysed for aluminium, carbon, chromium, copper, iron, lead, magnesium, manganese, silicon, sulphur, tin, and zinc. The components and materials analysed were:

pistons, transmission, heater and core, radiator, grille, trim, zinc diecast components, light gauge steel (3.2 mm), forged steel, cast iron, bumper, glass.

[23] Karl C. Dean and Joseph W. Sterner, *op. cit.*

Information for the fifteen individual cars was consolidated, in an unspecified manner, into metallic averages for a so-called 'composite' car; these averages formed the basis of the published work. While such averages are meaningless for individual vehicles, consolidation may be justified on the following grounds:

(a) Provided that the vehicle selection has been made on an accurate statistical basis, the composition of the composite car will indicate the tonnages of materials of given grade that should be available, nationally, from scrapyards.

(b) Scrapyards trade in a number of different grades of metal, each an aggregate of materials which, within limits, differ in composition. The copper-bearing radiator material used in one car may differ from that in another; all, however, are sold as 'radiator stock'. Product consolidation is, in fact, a routine operation in the scrapyard.

Wide variations may, however, be expected in hand-dismantling times from one model to another and this, as will be seen, may be significant in achieving low residual contents.

Dean and Sterner treated the residue, after stripping, as essentially ferrous. They took no account of, for example, lead as body filling. The only lead appearing in the analysis of the composite car was in the battery; some reports on individual cars list lead in wheel-balancing weights.

These reservations notwithstanding, the data of Dean and Sterner remain, despite their age, as the only readily available, detailed study of car composition, and have formed the basis for much of the discussion of automotive scrap both here and in the sources cited.[24]

It is important to note that gross revenue from secondary products is critically dependent on the grade that they command in the scrap market. For road vehicles, this revenue is heavily concentrated in the steel which, even if sold as a contaminated grade, would at prevailing metal prices have realised 43% of gross revenue in 1969, 60% in 1976, and 56% in 1986, reinforcing the historically pre-eminent position occupied by steel in the automotive scrap industry.

5.3 PRESENT RECOVERY PRACTICE

A discarded car usually goes to a salvager [wrecker] or dismantler, primarily a spare parts dealer who has a certain degree of flexibility in deciding the size of his stock of vehicles. This is likely to be determined by prevailing economic conditions, zoning restrictions, land values and litter laws.

---

[24] The work of Dean and Sterner contains certain minor arithmetical discrepancies, possibly due to rounding, that have been perpetuated by others, eg. Regan *et al*, and by Adams, who has introduced several additional but minor errors.

5.3.1 Stripping for saleable parts

For many years, certain areas of the scrap trade have specialised in the sale of used components which remain in serviceable condition in an otherwise unusable vehicle.

Long experience in the trade has shown that although it changes from one component to another the percentage of serviceable components remains more or less constant.[25]

Table 5.2
Percentage of reusable components in motor vehicles

Component	Percentage of reusable components
Engine	25
Axles, differentials and gears	25
Battery	20
Heater	10
Starter motor, generator	20

There has always been a demand for such parts from car owners who do their own repairs, from small garages who respond to a request from the customer to repair a vehicle without fitting a new factory component, or from the owners of vehicles for which new parts are no longer available.[26] The trade is not limited to mechanical parts; undamaged light fittings, bumpers, windows and panels from scrapped cars are all saleable. Parts from late models are in considerable demand; the price paid for a vehicle depends upon its age, model popularity and extent to which individual parts remain undamaged.

In recent years, dismantling for parts has lost some of its importance in countries where labour costs are high relative to the value of the serviceable parts. The problem is exacerbated in those areas where local demand for the *materials,* especially the recovered steel, is too low to provide an adequate financial return for the residue left after stripping for parts. Environmental considerations impose other constraints, such as the difficulty of disposing of seats, tyres and bulky wastes, such as carpets.

The extent of the use of salvaged components varies regionally, being greater in areas of low wages than in those of high. However, there are dual factors involved in the demand for old vehicles, viz. parts and scrap steel. High activity in the steel industry will create a high demand for

---

[25] Europool, *The disposal and recycling of scrap metal from cars and large domestic appliances,* London, Graham and Trotman Limited, 1978.
[26] Michael E. Henstock, The European Picture, *loc. cit.*

41

scrap, and enhanced prices will then make it worthwhile to collect over a wider radius. Since car breakers obtain most of their revenue from parts, the scrap value of stripped hulks is relatively unimportant to them and there is little incentive for them to attract the highest possible price by the most complete stripping to provide the cleanest steel.

The extent to which a dismantler strips a vehicle for parts will depend on its age and popularity. Some parts are still useful and profitable, and in marketing these used parts — whether for immediate use or for rebuilding [remanufacturing] — the dismantler helps to prolong the useful life of other vehicles. Ultimately, however, there comes a point at which the entire value of the vehicle resides in its recoverable scrap value.

After the removal of parts, the vehicle is normally stripped for non-ferrous metals. The battery may be sold for lead although, with the introduction of so-called 'strict product liability', it is becoming more difficult to find battery processors, especially in the U.S.A. Copper wire in; and small electrical parts will probably be left in the hulk when removal costs exceed their value; it has been estimated that, whilst 80% of radiators arriving at scrap yards may subsequently be recovered, only 50% of wiring, starter motors and dynamos or alternators are so removed. Engine blocks and other cast-iron parts are commonly removed and sold as foundry cast. Heavy steel components, including the chassis, often complete with front and rear-end assemblies, are then severed by cutting torch or shears, and sold to foundries as heavy melting steel.[27, 28] The hulk may then weigh less than 450 kg and seldom more than 900 kg, since 1/3–2/3 of the metal content of the car will have been removed. Problems arise with the remainder, which is commonly burned to remove non-metallics and then baled, sheared, or shredded, along with other light-gauge ferrous scrap.

Thus it may be seen that there are times when a combination of circumstances may make it unattractive to process obsolete road vehicles.

5.4   THE NATURE OF AUTOMOTIVE SCRAP

The dismantling data obtained by Dean and Sterner are very extensive, and only sections will be analysed here. Portions of the data, on U.S. cars manufactured *c.* 1960 and scrapped *c.* 1969, are presented in Table 5.3.

Ferrous prices are brokers' delivered prices; the figures quoted for No. 2 bundle steel are a composite of prices ruling in Chicago, Philadelphia and Pittsburgh, for the calendar year 1985.[29] Non-ferrous are dealers' buying prices, wholesale lots, f.o.b., quoted in *Recycling Today: Secondary Raw Materials,* Vol. 24, No. 12, Dec. 1986, as ruling in Dec. 1986.

The financial analysis is critically dependent on the grade that each recovered material can command.

[27] *Automobile disposal, a national problem.*
[28] Richard D. Burlingame, A scrap gap in the space age, *Scrap Age,* pp.45–49, June 1970.
[29] Staff, Inst. Scrap Iron and Steel, Inc., *Facts, 1985 Yearbook,* 43rd Edition, 1985.

Table 5.3
Recoverable metallic values of composite car[30]
(December 1986 metal prices)
Model Year *c*. 1960

Material	kg	Unit value ($/tonne)	Value ($)
No. 2 bundle iron[31]	1186·3	52	61·09
Cast iron[32]	194·7	90	17·52
Copper:			
Radiator stock	7·0	579	4.05
No. 2 heavy & wire	6·3	844	5·28
Yellow brass	1·2	728	0·75
Zinc, die castings	24·5	274	6·71
Aluminium, castings etc	23·0		12·52
Lead:		544	
Battery plates	9·1	62	0·56
Battery cable clamps	0·2	173	0·03
Total metallics	1452·3		108·51

5.4.1 Grades and prices

Grades for nominally identical grades of scrap can vary widely from city to city, especially in large countries such as the U.S.A., where transport distances can be considerable. The ruling price will depend on many factors, including:

(a) demand,
(b) supply,
(c) distance from market, and
(d) prevailing economic climate.

Thus, a certain degree of arbitrariness is inevitable in establishing average prices, since knowledge is needed of the quantities traded in individual cities on a particular day at a particular price. Even then, the very volatility of the scrap market renders such calculations unreliable except in the short term. With these *caveats*, the following grades and prices were assumed:

---

[30] Karl C. Dean and Joseph W. Sterner, *op. cit.*
[31] Based on total weights of metals recorded by Dean and Sterner. The analysis assumes that all light and heavy ferrous material, other than cast iron, will be sold as No. 2 bundles, rather than as No. 2 heavy melting scrap. It also assumes that 37·2 g of inseparable cast iron will be included with the No. 2 bundle material. The 'clean' steel separated by hand still analysed about 0·07% copper, i.e. it contained an additional 0·8 kg arising from this source; it is not included in the table.
[32] Does not include irrecoverable cast iron, which has been included with the No. 2 bundle material.

5.4.1.1 *No. 2 bundle steel (No. 5 bale)*

There are several categories of ferrous scrap in which automotive steel could, in principle, be placed. However, the attainment of a higher grade is usually a matter of confidence in the cleanliness of the material. Whether material is in fact pure is unimportant unless the purchaser is willing to believe in its purity and pay accordingly. To avoid arbitrariness in assigning a value to the proportion of metal placed in each category, the safest economic assumption, as well as the most realistic, was that all the steel would be sold as No. 2 bundle [No. 5 bale] grade. The price of $52/tonne was a composite price for 1985, the latest year for which complete figures were available.

5.4.1.2 *Cast iron*

Cast iron separated from non-ferrous contraries and from steel can be sold as clean automotive cast at a premium price in the U.S.A. Lower prices obtain for grades such as cupola or foundry cast. An average value of $90/tonne was assumed.

5.4.1.3 *Copper*

(a) *Radiator stock.* This includes the radiator and heater cores

(b) *No. 2 heavy copper and wire.* The remaining copper attains this grade. Any oxidation during incineration would not seriously upset the financial analysis.

(c) *Yellow brass.* The composite car contained about 1·2 kg of brass bushings, valves, sleeves and rivets, best sold as brass than for the 0·9 kg of copper that it contained. Prices varied in the range $460–660/tonne.

5.4.1.4 *Zinc*

Apart from the small amount in the yellow brass, recoverable zinc is present as die-cast material. Old die-cast scrap showed an even greater variation than did other constituents.

Unrecoverable die-cast and zinc used for rustproofing will largely report to the flue dust in the steelmaking operation.

5.4.1.5 *Aluminium*

Most of the aluminium is present in the cast form, for which there is a range of prices. A value of $544/tonne was assumed for this material.

5.4.1.6 *Lead*

Battery prices were $22–90 and soft lead $130–242/tonne. Values of $62 and $173/tonne were assumed.

5.4.2 Total financial value of scrap automobiles

Under the assumptions made, approximately $109 worth of secondary materials was theoretically recoverable from the composite 1960 model year vehicle. The time required for removal of components from the vehicle was 124 minutes, with a further 201 minutes for disassembly, a total of 325 minutes. The cost of stripping, at a nominal $10/h cost for labour, was $54.[33]

5.5   CHANGES IN THE MATERIALS USED IN VEHICLES

Changes in vehicle materials have been made for a variety of engineering reasons, including:

(a) cost,
(b) absolute performance,
(c) lightness, to improve fuel economy, and
(d) longevity.

A comprehensive review of the changes in automotive materials is not possible here. Recyclability, as it is discussed in these pages, is concerned largely with the economics of reclamation, and the effects of large-scale materials substitution. Some examples of these changes involve additional heavy metal trace elements, through the use of HSLA (high-strength, low-alloy) steels, additional phosphorus in specialised steels, more aluminium and plastics, and a reduction in zinc in trim, partially redressed by its increased use in anti-corrosion applications.

However, in addition to the large tonnage applications, such as the replacement of steel and cast iron by polymers and aluminium, new materials are finding their way into less-obtrusive applications. Engineering ceramics play an increasing part in automotive construction. Catalysts are supported on alumina ceramics, and zirconia/platinum composites find use in sensors, as do $Al_2O_3Cr_2O_3$ thermistors.[34] It is certain that scrap materials will in the future contain small concentrations of tramp elements which have not previously been encountered in them.

Much has been written about major changes to lighter materials, in order to enhance fuel economy. In the U.S.A. in particular, considerable effort has been devoted to substitution of steel and cast iron by aluminium and polymers.

Polymers are today used in automotive applications for a variety of financial and technical reasons. A particular advantage is that in applications where their properties meet the engineering criteria, they provide significant weight savings, and hence fuel economies, relative to metals.[35]

[33] The Annual Statistical Report of the American Iron and Steel Institute, 1984, gives the minimum standard hourly rate in the iron and steel industry, effective Feb.1 1984, as $9.90. Rates for 1976 and 1969 were $5.67 and $2.77/h, respectively. The corresponding publication for 1985, however, changes the basis of reporting, and gives total employment cost, including benefits, for wage-earning staff (as compared with salaried) as $21.29/h in 1984, and $22.81 in 1985. Unskilled labour rates (1987) in United States secondary metals industries are lower; they range from $5–7/h in non-unionised yards in the South to $12–15/h in unionised yards on the West Coast. Calculations in this study assume a rate of $10/h, for 1987, with corresponding figures of $5.75 and $2.75/h for 1976 and 1969, respectively.
[34] Anon, *New Materials and the Automobile,* Toyota Motor Corporation, undated.
[35] Anon, *The Energy Content of Plastics Articles,* Association of Plastics Manufacturers in Europe, Distributed in The United Kingdom by The British Plastics Federation, Publication No.309/1, April 1986.

Aluminium was successfully used in road vehicles at least as early as the 1930s, when one model of Riley made extensive use of it. In the 1950s, the Dynha-Panhard employed formed and spot-welded aluminium for the entire body structure, including the floor pan. Currently, the British Leyland Land Rover, which has an aluminium body structure, demonstrates exceptional longevity. Daimler-Benz, Porsche, Rolls-Royce and Volvo vehicles, especially those destined for use in the U.S.A., where Federal regulations provide for minimum values for fuel economy, have used aluminium in exterior body panels, doors, front wings, bonnet and boot lids. It is clear, though, that these vehicles are in the more expensive, or luxury end of the market. In cases such as Porsche, performance is the dominant reason to save weight. None of the applications is load-bearing, yet it is arguably in the load-bearing applications where aluminium alloys confer the greatest benefits in terms of weight-saving and durability.

One British vehicle manufacturer, BL (formerly British Leyland), through BL Technology Ltd, has carried out work with Alcan International Ltd on the development of weld-bonded aluminium structures for volume car production.[36] It has been established that such structures are limited by stiffness rather than by strength; hence, the use of readily available, medium-strength Al–Mg alloys has been shown to be possible.[37] The technology depends on the development of a coil product, which is pre-treated and pre-lubricated in coil form. Pre-treatment, lubricant and adhesive must all be compatible, so components can be press-formed from the coil product and weld-bonded without an intermediate cleaning stage.

The technicalities of the substitution of steel by aluminium in body panels, though not without interest, are not relevant to this study. So far as recyclability is concerned, the factors of importance are:

(a) Replacement of steel by aluminium, which will probably enhance the value of the hulk.

(b) Use of continuous adhesive bonding of flanges is not seen as deleterious to materials recovery or to the purity of the secondary metals so recovered, relative to the spot-welding alternative.

(c) Capacity or otherwise of the secondary aluminium industry to absorb increased quantities of scrap, which will be crucial to recovery activities.

The alloys which are proposed for use in the vehicle are the aluminium-magnesium alloys types AA 5251 (Al 1·7–2·4% Mg) and AA 5052 (Al 2.2–2.8 % Mg)

---

[36] M.J. Wheeler, P.G. Sheasby and D. Kewley, Aluminum Structured Vehicle Technology — A Comprehensive Approach to Vehicle Design and Manufacturing in Aluminum, *SAE Technical Paper Series*, No.870146, *Proc. International Congress and Exposition*, Detroit, Feb.23–27 1987.

[37] P.G. Sheasby, M.J. Wheeler and D. Kewley, *Aluminium structures in volume car production*, No publication details available.

It should be noted that design changes made for engineering or aesthetic purposes may or may not adversely affect recyclability. To the extent that they reduce materials values without a corresponding easing of recovery or reduction in recovery costs, recyclability will clearly suffer.

5.5.1 Changes in the materials balance in vehicles

For a variety of reasons, which normally devolve on cost, minor changes are continually being made in the materials balance of the vehicle. The use of plated polymers, often ABS, instead of plated zinc-based die castings in trim cannot necessarily be regarded as a permanent change, given the improvements in technology which continually give one a cost advantage relative to the other. The use of HSLA steels is undoubtedly permanent. However, so far as recyclability in its widest sense is concerned, the most important change is the reduction in the amount of steel used.

5.5.1.1 *Substitution of steel by aluminium and plastics*

The advantages of high payload/tare are obvious, and much may be done by improved engineering design. A point is reached, though, where further weight savings can be achieved only through the use of light materials; the magnitude of such fuel savings, over an estimated 174 000 km vehicle road life has been computed as 15–25 l/kg of weight saved.[38, 39]

The high energy investment required to produce primary aluminium may, it is suggested, be recouped by reduced fuel consumption associated with the lighter vehicles which make extensive use of the metal. Further, it may be remelted at an estimated energy cost of 2–10% of the original energy investment, depending on the form, and especially on the surface area/volume ratio, of the scrap aluminium. A further economy arises from the fact that aluminium need not be painted to confer corrosion resistance.

The principal scope for increasing aluminium use in road vehicles would seem to be in bodywork. As seen earlier, such practices are not new. The substitution is not, however, a straightforward operation. Joining presents problems. Aluminium panels can be at risk in markets such as the U.S.A. where hailstones can be large enough to dent them. One possible palliative for the latter problem is the use of age-hardening alloys. However, these introduce alloying additions which can themselves impair recyclability.

Current thinking seems to be moving towards the use of plastic panels, rather than aluminium, since the latter may present problems in repair. The principal objection to plastic exterior panels is their dimensional

[38] Kenneth D. Marshall, The economics of automotive weight reduction, Soc. of automot. Engrs, Paper No. 700 174, *Automot. Engng. Congr.*, Detroit, Jan. 12–16 1970.

[39] Marton C. Flemings, Kenneth B. Higbie and Donald J. McPherson, *Report of Conf. 'Energy conservation and recycling in the aluminum industry'*, Massachusetts Inst. of Technology, (co-sponsored by the Center for Materials Science and Engineering, M.I.T. and the U.S. Bureau of Mines, with the cooperation of the Aluminum Ass.), Jun. 18–20 1974.

instability and the difficulty of obtaining an acceptable appearance, especially when they carry attachments. Commercial vehicles, e.g. Post Office vans, are less critical in this regard. A detailed analysis of the substitution of steel and cast iron by aluminium and plastics lies outside the scope of this chapter. Substitution is carried out for economy in first cost or to obtain equivalent structural performance at lighter weight and, hence, with greater fuel economy. Some general observations are that the density and elastic modulus of steel are, respectively, 2·8 and 3·1 times that of aluminium. The highest strength-to-weight ratio of any material so far proposed for automotive use is that of glass-laminate epoxy. Glass fibre-reinforced polymers; have low elongations, of 5% or less. Iron and steel still have no real competition for high-strength, high-temperature applications.

Many components are simply covers, or serve cosmetic purposes. In the former category are bonnet and boot lids, and some wings. In these applications, gauge-for-gauge substitution by aluminium for steel is possible with weight-saving factors of 2·8. Double thicknesses of reinforced polymer could also be substituted, when the weight-saving factor for equal tensile strength would work out to a value of about 2.

5.5.2   The possible future composition of vehicles

The work of Dean and Sterner has been followed by other investigations into the materials balance of the vehicle, with particular reference to changes made in the interests of lightness or cheapness.

The energy crisis of 1973, with a resultant interest in better fuel economy has induced changes in purchasing patterns for cars. Although memories of fuel shortages have proved to be short, and large cars are again in demand, some of the move towards smaller vehicles is undoubtedly permanent. The future shape of the vehicle recycling industry may be analysed in terms of possible changes in size and weight of cars, and also in their composition and mode of manufacture. Smaller and lighter cars consume less fuel and materials, contribute less to congestion and pollution, and may be expected to be cheaper to purchase, maintain and operate. Weight reductions may occur by size reduction, and by substitution of aluminium and plastics for steel and, to some extent, for copper and zinc.

A detailed analysis by Roig *et al.* divided the car by size into three market categories for the U.S.A.[40] It presented data on 1975 model year vehicles and predicted those for the materials of 1980 and 1990 model year cars. Cars were classified as follows:

Class 'A' Full size family cars, six passengers
Class 'B' Compact cars, five passengers
Class 'C' Sub-compact cars, four passengers

[40] Robert W. Roig *et al.*, *loc. cit.*

The weights of these three groups are given as:

A >1590 kg
B 1135–1590 kg
C >1135 kg

Projections by the Aluminum Association of the penetration of the automotive market by aluminium suggested that changes up to the 1980 model year would be modest, with adherence to the basic steel structure. The most likely areas for substitution were in the so-called 'hang-on' parts, e.g. doors, bonnet and boot lid. The suggestion that an all-aluminium body would be a strong possibility by 1990 now seems unlikely to be realised.[41]

Roig has made forecasts on the basis of the following bases, termed scenarios:

(a) maximum credible aluminium
(b) maximum credible plastic
(c) most probable materials balance

The forecasts were made by considering item-by-item substitution possibilities, the corresponding weight savings, and an estimated fraction of production changeover. Consideration was taken of the compounding effects of weight saving, such that 1 kg saved in the upper structure of the car saves a further 0·5 kg in aggregate in components such as engine, transmission, chassis and brakes.

Scenario (c) only has been used in compiling Table 5.4. It was estimated that 80% of cast iron is replaceable by aluminium, and that 60% of this market could be penetrated by aluminium in scenario (a), and 20% for (b) and (c) by 1980. Under scenario (a), penetration would reach 100% by 1990; under (b) and (c) it would reach 40%.

(a) Assuming 30% Class A, 37% Class B, 33% Class C, including vehicles imported into the U.S.A.
(b) Including stainless steels
(c) Including malleable iron
(d) Including tin and magnesium. Alloying additions such as chromium, manganese, nickel, and tungsten are included in the steel weights.
(e) Total dry weights without fuel or fluids.

The work of Sterner *et al.* on Japanese cars should also be considered for the detailed materials breakdown that it provides on actual vehicles of relatively recent manufacture. It presents data for the composition of four models of Japanese car, three of 1981 and one of 1982 model year.

Examples of each model were hand-dismantled and the amount of each material determined. Stripping times were not reported. A simple

---

[41] C.N. Cochran, F.R. Abele, G.A. Alison, P.E. Anderson, L. Campbell, T. Eckert and F. Testin, Use of aluminum in automobiles — Effect on the energy dilemma, *Report of the Aluminum Ass. Task Force on Automotive Energy Saving*, Feb. 1 1975.

Table 5.4
Projected materials composition.
Most probable composition 1980 and 1990 (Roig et al.)

Material	Composite car (sales weighted)[a]		Composite car (sales weighted)[a]	
	1980		1990	
	(kg)	(wt.%)	(kg)	(wt.%)
Low carbon steel	643	50·4	530	46·3
Alloy steel[b]	83	6·5	91	7·9
Total steel	726	56·9	621	54·2
Cast iron[c]	174	13·6	91	7·9
Aluminium	81	6·3	136	11·9
Copper, brass	12	1·0	6	0·6
Zinc	5	0·4	4	0·3
Lead	10	0·8	8	0·7
Other metal[d]	9	0·7	16	1·4
Rubber	65	5·1	58	5·0
Glass	34	2·6	32	2·8
Polymers	85	6·7	105	9·2
Other non-metal	76	5·9	68	6·0
Totals[e]	1277	100·0	1145	100·0

average, with no attempt to apply sales-weighted corrections, yielded the following weights of materials:

Quite clearly, the materials value of the average Japanese car, to the extent that it can be determined from four individual vehicles, is very much lower than the $108·51 of the composite vehicle of U.S.A. manufacture some twenty years earlier (Table 5.3). However, the important factor from the viewpoint of the materials reclaimer is the value of the *recoverable* materials contained in the entire input to dismantling or scrap processing yards. For such purposes, a sales-weighted analysis, such as that of Roig et al., will give more representative figures.

Examples of each of the four Japanese cars were also shredded, and the weights of each of the recovered materials were determined (Table 5.6).

The investigation also included a study of the HSLA (high-strength, low-alloy steels); which comprised 12% of the total ferrous metals in one of the models; HSLA steels are now extensively used in vehicles worldwide, and have always been recognised as presenting a potential recycling problem. Total separation of an HSLA steel product by hand-dismantling was not considered feasible. Therefore, a programme of melting and analysis of the ferrous metals was carried out to determine whether the HSLA would introduce undesirable tramp elements to an unacceptable extent. It was concluded that, at current alloying levels, ferrous scrap

Table 5.5.
Average recoverable materials and values in Japanese cars[42] (1986 metal prices)

Material	Weight (kg)	Value ($)
No. 2 bundle steel	612	31·82
Cast iron	56	5·04
Copper:		
Radiator stock	8·6	4·98
No. 2 heavy & wire	2·7	2·28
Zinc die castings	2·4	0·66
Aluminium	38·4	20·89
Lead (ex. batteries)	0·2	0·03
Total	720·3	65·70

Table 5.6.
Materials recovered in shredding and water classification of Japanese cars (Average of four cars)[43] (1986 metal prices)

Material	Weight (kg)	Value ($)
No. 2 bundle steel	538	27·98
Cast iron	53	4·77
Copper:		
Radiator stock	6·8	3·94
No. 2 heavy & wire	1·9	1·60
Zinc die castings	2·1	1·20
Aluminium	25·4	13·82
Lead (ex. batteries)	0·2	0·02
Total		53·33

containing them would not present a problem with current recycling technology.

5.6 THE EFFECTS OF MATERIALS SUBSTITUTION ON VEHICLE RECYCLING

At present, and for the foreseeable future, the primary product of the automotive recycling industry is ferrous scrap, and at least 50% of the

[42] J.W. Sterner, D.K. Steele and M.B. Shirts, *Hand dismantling and shredding of Japanese automobiles to determine material contents and metal recoveries,* Washington, D.C., U.S.Department of the Interior, Bureau of Mines, RI 8855, 1984.
[43] *ibid.*

average vehicle will be ferrous until 1990 under any of the projected compositional scenarios.

Any reduction in steel usage in vehicles will affect the economics of dismantling and processing them. A reduction wholly in favour of plastics, at present essentially valueless when recovered from a hulk, would reduce overall vehicle value. However, replacement by aluminium could, by 1990, have a counterbalancing effect.

Significant conclusions regarding changes in recoverable values are difficult to make since no data are available that are precisely comparable with those of Dean and Sterner for the 1960 composite car. However, the nominal compositions derived by Roig et al. for a typical, sales-weighted 1975 model sold in the U.S. (Table 5.4), and projections for other years, may be used, with scrap metal values as quoted in Section 5.4, to determine the relative value of the recoverable materials contained in

Table 5.7.
Changes in scrap values in 1975 composite car and in 1980 and 1990 model years, relative to 1960 (1976 and 1986 metal prices)

Material	1975		1980		1990	
	a,b Change (wt) kg	c,d Change (value) $	a,b Change (wt) kg	c,d Change (value) $	a,b Change (wt) kg	c,d Change (value) $
Steel	−233	−12·11	−460	−23·92	−565	−29·38
Cast iron	+58	+5·22	−19	−1·71	−102	−9·18
Aluminium	+22	+11·97	+58	+31·55	+113	+61·47
Red metals[e]	+0·5	+0·33	−3	−1·95	−8	−5·2
Zinc	−11	−3·01	−19	−5·21	−21	−5·76
Lead	–	–	–	–	−1	−0·06
Net change relative to 1960*		+2·4		−1·24		+11·89
Net changes relative to 1960*		−0·86		−11·19		−9·55

a   Data for 1975 vehicle and projections for 1980 and 1990 are from Roig, *op. cit.*
b   Reference points are the data of Dean & Sterner, *op. cit.*
c   Other metals (alloying elements etc. and non-metallics are assigned zero value.
d   Metallic values as in Table 5.3 and Section 5.4.1, for 1986, and from a similar exercise for 1976.
e   Red metals: copper and its alloys.
*   Calculated at Dec. 1986 prices
** Calculated at Apr. 1976 prices

1960, 1975, 1980 and 1990 model year vehicles. These changes are given in Table 5.7.

It should be noted that:

(a) The profitability of stripping cars is very dependent on scrap prices. Since these can change significantly, particularly in the short term, there are correspondingly large changes in profitability.

(b) A projected decline in steel and cast iron in cars between 1960 and 1990 model years is reflected in a decline in the calculated dollar value of hulks between these model years of about 10% if calculated at 1976 metal prices. The application of 1986 metal prices, which show an increase in the value of secondary aluminium between 1976 ($353/tonne) and 1986 ($554/tonne), has reversed this decline in hulk dollar values. This change in the value of the recoverable scrap aluminium has been one important factor — another is the introduction of the shredder — in maintaining the profitability of the recovery operation. It follows that a fall in the unit value of secondary aluminium could be equally significant in eroding that profitability.

(c) Fig. 2 shows that metals which would in 1976 have commanded $106.85 would, by 1986, have commanded only $108.51, an increase in ten years of 1.6% of the 1976 value. The cost of the necessary labour for hand-dismantling; would, however, have risen from $31 in 1976 to $54 in 1986, a rise of 74% on the 1976 costs in the same period. No realistic data are available for the cost of disposal for non-metallic detritus in 1976. It will be seen later [Table 5.8] that, assuming no other changes, movements in metal prices alone during 1976–1986 have reduced margins from $36.54 to $7.23 for those components which, as will be seen, provide the most financially attractive areas to the dismantler. No data are available for changes in possible labour requirements.

Clearly, by suitable choice of date, the analysis could be made to yield very different answers. However, it is clear that labour costs have increased at a much faster rate than have the revenues from the recovered metals.

5.7 THE PROFITABILITY OF STRIPPING AND BALING

Assuming a basic labour rate of $3/h and a 20% overhead, Dean and Sterner concluded that a hulk could be bought for $9, stripped of copper-containing parts to achieve a copper content of <0·1%, baled, and could earn a 19% DFC rate of return before tax. By designing a baling operation and estimating costs, they hulk to a bundle [bale], including depreciation of the smokeless incinerator used to eliminate the non-metallic combustible materials.[44] The estimated revenue was $55.94/

---

[44] Doubts have been expressed concerning the economics of the Bureau-designed incinerator, whose low capital cost partly derives from the employment of second-hand constructional materials.

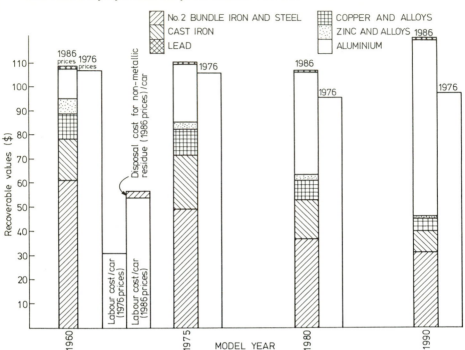

Fig. 2 **Changes in recoverable metal values and in labour costs for vehicles of 1960–1999 model year at April 1976 and December 1986 metal prices**

hulk. This indicated that, at 1969 prices, a high-quality bundle could be produced from a *c.* 1960 model car at a cost that permitted a reasonable price to be paid to dismantlers or last owners.

Other workers expressed the view that, although many mills still purchased well-prepared bundles, demand was so depressed that, with existing costs and revenues (1970), complete hand-stripping to the extent necessary to produce a bundle of acceptable quality was not viable.[45]

One possible reason for this discrepancy may lie in the assumption that the purchased vehicle is complete on receipt, and that it is then dismantled to the extent suggested by Dean, i.e. complete disassembly. Examination of the structure of the vehicle scrap industry shows this assumption to be unrealistic. The briefest summary of the dismantling industry is as follows:

(a) Typically, an obsolete vehicle is first stripped of saleable parts and of any easily-removable, high-grade metals whose value exceeds the marginal cost of removal and sale.

[45] Richard D. Burlingame, *loc. cit.*

54

(b) The hulk is burned to remove non-metallic residues.

(c) Further hand-stripping of valuable non-ferrous metals is followed by baling, shearing or shredding; this last will be followed by magnetic separation of ferrous from non-ferrous metals.

Early designs of shredder did not readily deal with engines, transmissions, and other assemblies containing hardened steel components, and prior removal of these items is still the preferred practice. Many installations stipulate removal of tyres and seats before shredding, and all insist on the removal of the petrol tank. By contrast, the baler can accept complete vehicles. It is unlikely that a dismantler will strip vehicles to two different specifications, viz. those of a shredder or of a baler operator, when in the volatile secondary metals industry, the relative advantage of selling hulks to one or the other can change from day to day. Stripping is more likely to be standardised to the more stringent specification, i.e. that of the shredder, so that hulks may be sold to either installation. In fact, the willingness or otherwise of scrap processors to accept them depends entirely on current conditions in the scrap markets. In times of scrap shortage and high prices, processors will accept vehicles in an incompletely stripped condition that they would deem unacceptable under conditions of greater scrap supply.

5.8 THE POSSIBILITIES FOR SELECTIVE STRIPPING

Quite apart from the requirements of the baler or shredder operator, a more direct financial stimulus for prior removal of certain components may be found in an analysis of the marginal costs and revenues of stripping.

The hand-stripping data of Dean and Sterner have been divided by Adams into five groups:[46]

Group A: Includes battery, radiator, engine, transmission, flywheel housing, starter and solenoid, dynamo (generator), distributor, coil, thermostat, fuel pump, carburetter, engine wiring.

Group B: Includes differential, heater core, body wiring, wheel covers and wheels.

Group C: Includes voltage regulator, electric motors, dashboard wiring, instruments, radio, interior trim, mirrors and light covers, interior and exterior handles, horn and relay, tubing, exterior trim and grille, window frames, brake drums and cylinders, bumpers and trim, steering gear box.

Group D: Heavy and light iron, *i.e.* No. 2 bundle material.

Group E: Rubber, glass, non-combustibles.

[46] Robert Louis Adams, *op. cit.*

Application of the dismantling data of Dean and Sterner, and calculation at 1976 and 1986 metal prices, yields Table 5.8.

The total revenues from the sale of Groups A–C Adams termed the 'Total non-No. 2 bundle scrap revenue', and on this basis, and on the detailed dismantling data of Dean and Sterner, he determined the contribution of each group to the marginal cost and revenue, also shown in Table 5.8 for 1976 and 1986 metal prices and labour costs.

Table 5.8
Marginal costs and revenues of selective hand-stripping
(1976 and 1986 metal prices and labour costs)

Group	Marginal revenue from groups A, B and C				Marginal cost of stripping groups A, B and C		
	($)		(%)		($)		(%)
	**	*	**	*	**	*	
A	37·18	38·66	65·4	63·0	7·92	21·08	39
B	9·36	8·59	16·4	14·0	2·26	6·00	11
C	10·34	14·16	18·2	23·0	10·16	27·10	50
Total	56·88	61·41	100·0	100·0	20·34	54·18	100

** 1976 metal prices and labour costs.
 * 1986 metal prices and labour costs.

Table 5.8 makes clear the erosion of the dismantler's margin; in 1976, the margin between revenues and raw labour costs for dismantling the composite car was $36.54. By 1986 that margin had shrunk to $7.23. Neither of these two figures takes into account the cost of capital equipment, utilities, or any of the other fixed or variable costs of a scrapyard.

It is therefore clear that the dismantler can remain viable only by a modification of the full dismantling procedure. The components listed in Group A produce, at 1986 metal prices, 63% of the 'Non-No. 2 bundle revenue' attainable from complete hand-stripping, but incur only 39% of the marginal stripping costs. Furthermore, all Group A components are associated with only four main items, i.e. engine, transmission, radiator and battery. By preparation of these four items alone, not including their associated equipment, $30.69 of a possible $56.88, i.e. 54% (at 1976 prices and costs) of the possible yield could be realised. At 1986 prices, these four items would yield $34.95 of a possible $61.41, i.e. 57% of the non-No. 2 bundle revenue'. By contrast, at 1976 prices Groups B and C incur 61% of the hand-stripping costs and recoup only 35% of the non-No. 2 bundle revenue. Thus, even when the dismantler *who receives a complete hulk* does not require these components as spares, their metallic content might make it worthwhile to remove them prior to sending the hulk to the scrap processor.

Traditionally, the scrap processor has received most of his hulks from dismantlers. Additional hulks are purchased from independent collectors, who are equally likely to be attracted by the profitability of these four items.

Thus, Adams argued, it is likely that the typical hulk received at a baling facility will lack the profitable Group A components. Since Groups B and C are commercially unattractive, one might then expect that the hulk would be incinerated and baled without further work.

Such reasoning, however, ignores the fact that marginal revenues from Groups A–C include sums notionally — but probably not actually — arising from the category designated 'Other ferrous', and assumed to be sold as No. 2 bundle material. To the extent that Groups A–C include such material, their financial value is artificially inflated and that of Group D reduced. Therefore, the validity of the argument must be examined in the light of prevailing market conditions for No. 2 bundle material.

(a) If the value of No. 2 bundles is so depressed that the value of No. 2 steel contained in a hulk would not repay the cost of transport, it is essentially of zero value. The revenue from selective stripping would therefore form the entire income from the hulk; that revenue would, however, be reduced by the value of its No. 2 bundle component. In such circumstances, it would be worthwhile to remove selected components and, subject to disposal costs, to abandon the remainder; recalculated values are given in Table 5.9.

It can be seen that Group A now assumes even greater relative importance, at 70% of the effective revenue (75% at 1976 prices), especially since total revenue is concurrently reduced from $61.41 to $47.18, i.e. by 23·2%.

(b) There can, though, be few periods when scrap prices remain depressed for long enough for a No. 2 bundle to have no market for

Table 5.9
The profitability of selective stripping, assigning zero value to the No. 2 bundle scrap (1986 metal prices).

Group	Marginal revenue ($) (I)	No. 2 bundle component revenue ($) (II)	Remaining revenue ($) (I)–(II)	Percentage of total remaining (%)	Percentage of marginal cost (%)
A	38·66	5·43	33·23	70·4	39
B	8·59	4·27	4·32	9·2	11
C	14·16	4·53	9·63	20·4	50
Total	61·41	14·23	47·18	100·0	100

the foreseeable future. Markets usually exist even if, and some-times because, prices are low. Although most scrap processors experience periods of slack demand, they nonetheless often take advantage of the resulting low prices to build up their stocks.

At times when the No. 2 bundle has a value, removal of selected parts is proportionately less important to the dismantler, and it is appropriate to consider the total value of the hulk, as given in Table 5.10.

Table 5.10
The contribution of grouped components to total recoverable value (1986 prices)

Group	Marginal revenue ($)	Percentage of total revenue (%)	Percentage of marginal cost (%)
A	38·66	35·5	39
B	8·59	7·9	11
C	14·16	13·0	50
D	47·43	43·6	–
Total	108·84	100·0	100·0

It is not necessary to apportion marginal costs to Group D since, after necessarily incurring costs in removing Groups A–C, the Group D resi-due might be considered to be obtained gratis. In fact, some stripping or burning of non-metallics, which incurs cost, is normally necessary before sale.

Note that one cannot obtain the marginal revenue of Group D without first removing Groups B and C, each of which produces

5.8.1 Influence of selective stripping on the levelof residual copper in automotive steel scrap

Group A items involve 9·7 kg, or 68% of the total copper in the vehicle. After their removal, a residue of 4·6 kg is left and the body steel itself may contain 0·8 kg. Thus, 5·4 kg of copper remains in a hulk weighing 1145 kg (112·3 + 120·5 + 912·2 kg), representing 0·47% Cu, assuming that most of the non-metallics of Group E are eliminated in combustion. This content is very close to the average copper content of 0·48% reported by Battelle for No. 2 bundles.[47]

It is obviously uneconomic to recover Group C items for their marginal revenue, but Group B could be profitable. Removal of this group would reduce the remaining copper contamination by 3·0 kg, and the copper content of the remaining steel to about 0·23% Cu, i.e. approximately that

[47] Battelle Memorial Inst., *The measurement and improvement of scrap quality,* Report to the Institute of Scrap Iron and Steel, Inc., Columbus, Battelle Memorial Inst., Dec. 23 1960.

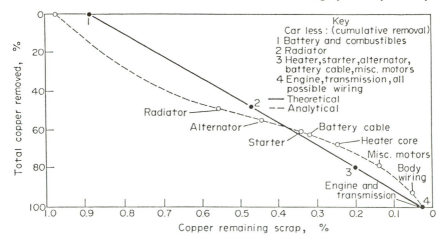

Fig. 3   After U.S. Bureau of Mines

of shredded scrap. The bundle might then attract a premium price from a steelmaker who knew the stripping to have performed conscientiously.

By a programme of melting weighted proportions of components, based on materials distribution, Dean and Sterner assessed the progressive reduction of copper content, during hand-stripping. The effect of stripping, in order of ease of removal, is shown in Fig. 3 for a 1963 Dodge.[48] The analysis of the hulk presented difficulties in representative sampling, but showed acceptable agreement with theory.

A further analysis of the removal of copper from cars is due to Stone, and remains unpublished. It predated publication of the dismantling data of Dean and Sterner, but used some of their results. Its aims were to:

(a) Identify copper sources in vehicles.

(b) Classify the location and attachment of copper-containing components.

(c) Evaluate contemporary salvage and scrapping procedures in terms of their efficiency of copper removal.

(d) Analyse the economic factors involved in the production of low copper-content vehicular scrap.

(e) Examine the possibilities for the redesign of automotive components to eliminate copper or reduce its level providing for easier maintenance.

The work of Stone illustrates that typical salvage operations, carried out on six of the fifteen cars which comprised the composite vehicle of

---

[48] *Automobile disposal, a national problem.*

Dean and Sterner, removed components containing some 73–89% of the copper in the car, either for resale parts, for metallic value per se, or because they were attached to larger components which were normally removed. Thus, the dismantling operations typically left in the hulk between 12 and 27% of the original copper.

The other important parameter is that of component removal time. Here, the analysis presented by Stone is based on:

(a) The dismantling data of Dean and Sterner,[49] and
(b) Dismantling times recorded in 20 different yards over a period of 60 days.

The latter seem more representative of removal times habitually incurred in the industrial environment by experienced operatives who are necessarily more concerned with yield of metal per unit time than with precise scientific measurement.

Material from the two sources was consolidated by Stone in an unspecified manner. The final data show a marked spread in the reported time required to remove components. This is attributable to differences in product manufacturing method and in dismantling procedure. Most components were removed by semi-skilled labour, with cutting torch or simple handtools, particularly in yards which received and processed only a few cars daily, and could not justify complex and expensive machinery or procedures. However, the improvements made by modifications such as assembly-line operations, properly designed and sized tools, and by trained personnel, could be considerable, especially in yards which specialised in scrap metal rather than in used parts.

Stone's data also exhibit differences from those of Dean and Sterner; in about 50% of individual removal operations the data established by Dean fell within the spread claimed by Stone. Where they did not, Dean's reported removal times were generally the longer, most noticeably in the case of the body wiring, i.e. the wiring associated with tail-light area, interior lights and other electrical items not associated with the dashboard, e.g. cigarette lighter, electric seat and window motors.

It was claimed that, on average, only 10–12 minutes was required to remove approximately 88% of the detachable copper present in the car, since most of it was concentrated in the radiator, dynamo, starter, battery cable, and heater components. Further tests showed that removal time for the wiring and miscellaneous components (relays, regulator, instruments, horn, radio etc.), comprising the other 12% of the copper, ranged from 4·2 to 20 minutes, with an average of 12. These times could probably be reduced if the operations were performed routinely.

The wiring left in the hulk contributed 0·17–0·27% Cu to the steel. Satisfactory and economical methods of wire removal from behind the dashboard and from the body had not, at that time, been developed, and

---

[49] Karl C. Dean and Joseph W. Sterner, *op. cit.*

are not known to exist today. The problem was greatest when incineration could not be performed, since burning removed insulation, upholstery and floor mats, and exposed much of the wiring for subsequent hand-stripping. The task was regarded by yards as an imposition, and it was carried out only when a particular customer paid a premium for the scrap. At the time, one particular scrapyard offered no fewer than seven grades of automotive slab, depending on the degree of non-ferrous removal. The lowest grade (Dirty) had been incinerated. The highest grade (Super) was virtually free of non-ferrous material. The price differential slightly exceeded $9/tonne, which should be compared with a yearly composite price (1968) of $29/tonne for No. 1 heavy melting steel — a superior grade — or $22/tonne for No. 2 bundles.

The provision of the manpower and quality control necessary to attain low copper levels could be justified, financially, if there were a premium for low-copper scrap. This is especially unlikely in the case of the smaller or less-favourably located yards, which have no contract to supply guaranteed quality scrap and must compete in the open market, where the bundle and automotive slab are treated as low-quality steel, and command no premium.

Dean and Sterner's data for the radiator, dynamo, starter, battery cable and heater core (but not the heater motor, which is not individually identified) indicated that these parts contained 9·9 kg copper, or about 69% of the detachable metal, and that their removal required 12·5 minutes. A further 1·8 kg resided in the body wiring, and required 16·6 min for removal, a total of 29·1 min. If, however, in respect of this item alone, the (more likely) time determined by Stone (4 min) is accepted, 82% of the detachable copper may be eliminated in about 16 minutes, as compared with the figure of 88% in up to 12 minutes, quoted by Stone.

5.8.1.1 *Significance of dismantling data*

Only limited data correlate the copper content of bales (bundles) with the extent to which they include automotive scrap, and they suggest a higher-than-average copper content in those derived largely from vehicles. Analysis of bales produced entirely from vehicles in the U.K., gave an average copper content of 0.45%.[50]

Dismantling practice and efficiency undoubtedly vary considerably from one yard to another and with changes in metal prices, and only limited conclusions may be drawn from the dismantling data available:

(a) If 0·44% is representative in copper in a No. 2 bundle, and if, as observed by Stone, 75–90% of the 1% Cu typically contained in a car is removed before baling, it implies that bundles contain some non-automotive scrap.

(b) Cars stripped for immediate gain, i.e. for removal of valuable copper-bearing components but not of the body wiring would, if

[50] M.W. Hubbard, The use of low grade scraps. Part I. Light gauge wrought iron and steel, *J. of Research*, Steel Castings Research and Trade Association (SCRATA), (32), pp.14–22, Mar. 1975.

isolated from the other scrap, and incinerated and baled, contain a total of about 0·36% Cu.

A computer analysis of car dismantling has been made by Sawyer, on the basis of a model that treats the car as consisting of 40 distinct parts, i.e. the 36 listed by Dean and Sterner, plus tyres, windows, seats and petrol tank, all of which, it was assumed, were removed prior to shredding.[51] A number of possibilities were considered for each component:

(a) Whether or not to remove it.
(b) Whether to sell a removed component *in toto*.
(c) Whether to dismantle a component for metals, or
(d) to dispose of the component as solid waste.

Application to the model of then current labour rates ($6/h), materials prices and cost of disposal of the residuals, yielded an optimum residual level of 0·49% for copper and 0·05% for tin remaining in the hulk; the value given by Battelle for copper in No. 2 bundles, 0·48%, is in good agreement. The Battelle figure of 0·08% tin shows only fair agreement.

The full assumptions regarding the sale of products are not stated in the analysis, but it is not surprising, given any reasonable set of metallic values, that the dismantling model has derived an optimum stripping level consistent with the removal of Group A items, where valuable cast iron and non-ferrous metals are heavily concentrated in a few components. Thereafter, the model is less successful; it predicts that, relative to optimal stripping as above, the cost of further copper removal to a representative shredder quality (0·22%), was about $0·26 for a hulk weighing 1 tonne. This implies, at its assumed $6/h, an additional 2.6 man minutes and, from the complete dismantling data, it is clear that very few components, and none containing a significant amount of copper, is detachable in so short a time.

The model makes no provision for an increase in value of the residual steel with progressive copper removal.

5.8.2 Salvaging costs

Stone stated that only 10–12 minutes, excluding internal yard transport of hulk and products, were required to remove the significant components, normally stripped in salvage yards, which contain approximately 88% of the copper.[52] On average, it required an additional 12 man minutes to remove the remaining copper-bearing components. At a notional hourly rate of $10/h the average dismantler could, on this basis, essentially strip the car of all copper-bearing components for a labour cost of around $4. Allowing an additional expense of $2 for labour to transport the components about the yard, and to store them, Stone's dismantling data suggest a total of approximately $6 as a reasonable

[51] James W. Sawyer, Jr., *op. cit.*
[52] Stone, *op. cit.*, p.4 states 12 minutes; p.17 gives less than 10 minutes.

estimate for the dismantler's complete costs for the 24 man minutes needed to achieve copper removal of 98% or better.

Since the actual removal of salvageable components normally takes only about half this time, it appears that a demand for complete copper removal would require an additional payment to the dismantler of about $2–3 per vehicle, more or less, depending on prevailing labour rates.

The price differential between No. 2 heavy melting steel and No. 2 bundles in April 1976 varied from $5 to $30/tonne. In December 1986 it was in the range $6–23/tonne. Assuming, for illustration, an average premium for No. 2 heavy melting of $15/tonne over No. 2 bundles, the 1·186 tonne of ferrous material assumed to be of No. 2 bundle grade, and valued at $65.25, would command $83.04, i.e. a premium of almost $18. At a notional hourly rate for 1986 of $10.75, the additional labour that such a premium could support would easily cover hand-stripping for complete copper removal. In fact, the detailed dismantling data of Dean and Sterner show that the composite car of their investigation could be completely dismantled to its component materials in about five and one half hours.

As shown above, if the achievement of low levels of residual copper is the aim, there is in fact no need to perform complete disassembly. The crucial factor is the establishment of the trust and confidence of a buyer in the scrap processor and in the consistent quality of the product, such that the premium price will be paid.

5.9 RECOVERY OF NON-FERROUS METALS FROM VEHICLES

Apart from the hand-picking of larger pieces of non-ferrous metal, iron and steel were the only products in the early days of the scrap shredder. Now, however, considerable effort is directed towards the recovery of the non-ferrous metals left behind after magnetic separation of the steel and cast iron. Methods used include:

Air and water elutriation, separation by trajectory, ferrofluids, heavy medium, cyclones operating on water or heavy–medium cyclones, cryogenics, sweating, electrostatics, and electromagnetics. A review of European practice in recovery of non-ferrous metals from shredder residues is due to Rousseau and Melin.[53]

5.10 BARRIERS TO VEHICLE RECYCLING

The barriers to vehicle recycling are simple. Briefly, they are that when a vehicle contains parts or materials which are of too small a value to justify separation for sale under prevailing market conditions and environmental constraints, there is no financial incentive to reclaim them. Currently, the trend in materials use in vehicles is towards the employment of smaller quantities of recoverable and saleable material

---

[53] M. Rousseau and A. Melin, The Processing of Non-Magnetic Fractions from Shredded Scrap, to be published in *Resources, Conservation and Recycling*, 1988.

and larger quantities of materials whose disposal incurs a cost. The cost of the labour needed to recover the materials of the vehicle has, over a ten year period, increased much more rapidly than have the revenues from the operation.

5.11 POSSIBLE VEHICLE DESIGN CHANGES TO FACILITATE RECOVERY

Few, if any, yards can operate on the scale of the hypothetical one proposed by Dean and Sterner. Most dismantlers are independent of smelters and operate on a small scale. They confine themselves to the removal of easily marketable components such as battery, radiator, engine and ancillaries, and sometimes the radio. However, the copper residing in components which provide insufficient marginal revenue to justify their removal, still exceeds the amount deemed acceptable by the steelmakers.

Because of the serious effect that it has on the desirability and hence the price of baled car scrap, much of the preceding discussion has been in the context of copper contamination in steel. There is, of course, also an effect on shredded scrap, but the copper content of this is lower and also less variable. While complete separation, i.e. zero contamination, is always the aim, and will maximise revenue from the sale of products, some impurities are less objectionable than others and need not be removed to the same degree of completeness. There is, therefore, scope for substitution. However, materials substitution is thought to be much less likely to happen for the reasons of recyclability than for broader economic or financial reasons.

Contamination might be avoided in the following ways:

(a) Changes in design

- Mechanical disassembly might be simplified.
- There might be an effort to avoid self-contaminating *combinations* of materials.

(b) Changes in materials

- Materials might be standardised.
- Materials might be identifiable.
- Harmless materials might replace deleterious ones.

(c) Changes in reclamation techniques.

Areas (a) and (b) are of relevance to this study. They are the responsibility of the vehicle manufacturer.

5.11.1 The role of the vehicle manufacturer

Clearly, the manufacturer will, voluntarily, implement design or manufacturing changes only if they involve no cost increase. The manufacturer cannot be blamed for this. He is, in fact, responding to a consumer who, while willing to pay more for some features than for others, is notably influenced by first cost. A few purchasers seem, for example, to be prepared to pay more for enhanced vehicle life. Naturally, a car manufacturer — who must survive in a highly competitive market — cannot afford to incorporate design changes for recyclability if such changes increase the cost and the product becomes uncompetitive.

Design changes made for improved recyclability will not necessarily increase manufacturing costs. However, in an industry which is as cost-conscious as is the automotive, present manufacturing methods are presumably the most cost-effective ones under current conditions. Design for recyclability is, therefore, likely to increase costs.

Only limited evidence suggests that purchasers are willing to pay a premium for enhanced vehicle life.[54] There is, however, no evidence at all that they are willing to pay for enhanced recyclability. Cost increases associated with changes made unilaterally in order to ease recycling would place any one manufacturer at a competitive disadvantage. To this extent, design for recyclability is likely to be considered only if essential for some other reason. However, the political and economic implications of legislation to encourage or compel manufacturers to use higher proportions of secondary materials, or of designing for recyclability, are far-reaching.[55, 56, 57]

Vehicle manufacturers claim that their products are almost completely recycled, and have been for many years. The accumulations of derelict vehicles at times of weak demand for scrap are, however, evidence to the contrary. While some major companies have encouraged attempts to ensure that cars are more quickly and easily recycled, this concern does not yet seem to have been translated into action at the design stage.

With large capital investment in plant that may be incapable of implementing radical design changes, it was to be expected that manufacturers would direct their efforts less towards redesign than towards the development of recycling techniques for existing models. Both Ford and General Motors have provided the United States Bureau of Mines with late model cars for investigation of recycling parameters. No large vehicle manufacturer, though, appears especially interested in taking recyclability into consideration at the design stage. Recyclability appears to rank very low on the list of reasons for design change, although consideration has been given to the feasibility of making valuable components, such as dynamos, more easily removable from the car so that they could be removed prior to shredding. However, they then become easier to steal and this is

[54] B.B. Hundy, The durability of automobiles, *Resources Policy,* **2,** (3), pp.179–192, Sept. 1976.

[55] Michael E. Henstock, Realities of recycling, *Symp: Recent advances in the recovery of useful materials from industrial waste,* London, The City University, (Mar.2 1976), *Chemy Ind.,* (17), pp.709–713, Sept. 4 1976.

[56] Anon., *Second Report to Congress, Resource recovery and source reduction,* Washington, D.C., United States Environmental Protection Agency, (SW-122), Mar. 26 1974.

[57] See, for example, Bert-Axel Szelinski, The New Waste Avoidance and Waste Management Act (WMA), and C. Grignaschi, Plastics Recycling in Italy: Legislation, Industrial Experiences and Future Developments, both to be published in *Resources, Conservation and Recycling,* Vol.1, 1988.

evidently a major problem, especially in large cities, where cars may be stripped of valuable components in a matter of minutes. The problem of theft has, in at least one case, resulted in design changes directly counter to recyclability, when factory-fitted stereo music systems are built in, in what is termed an 'integrated' design that makes theft difficult. The economics of integrated and non-integrated construction have not been published.

While it is felt that design for recyclability lies well in the future, it is possible to contemplate improvements that result from changes made for other reasons.

5.11.2   The objectives of design for copper removal

The principal impediment to the use of automotive-derived steel scrap has been shown to be copper, which is more troublesome in baled than in shredded scrap. It is true that the baler has lost much ground to the shredder, which produces scrap of lower copper concentration. However, even shredded scrap still contains approximately 0·22% Cu.

Since scrap is a recirculating load, and will pick up contamination on each recycling, each successive cycle will cumulatively raise the impurity level. That level would, therefore, rise inexorably in steel produced from scrap were it not for the fact that there is continual dilution with new steel, which is generally of low residual level. Although there is no immediate sign of a shortage of primary iron it may be that future shortages of coking coal, of natural gas, which is used in Direct Reduction plants, or of energy generally, will make it less desirable or financially less feasible, to use new iron. By contrast, since steel production from scrap has many economic advantages, its wider use is desirable. Hence, it is important that the secondary raw materials, including scrap, should be of as high a purity as possible.

The maximum acceptable level of copper in scrap for steelmaking use is about 0·15% and, since the steel may already contain 0·07–0·09% in solid solution, only 0·06–0·08% extra is permissible. Elimination of all detachable copper except the body wiring would leave 1·8 kg in a residual ferrous hulk weighing 900–1200 kg, depending on whether the stripping of copper-bearing components were followed by recovery of their ferrous metal. Thus, the residual steel would contain 0·15–0·20% Cu, which is very much on the border line of acceptability for deep-drawing applications.

The total amount of copper used in cars remained almost constant during the period 1965–1975; increased efficiency of copper use in, for example, radiators in U.S. cars was counterbalanced by an increase in the incidence of components previously considered as 'luxury'. Such components include air-conditioning (from 23·3 to 72·6%), power steering (from 53·8 to 89·9%), power seats and windows, radio aerials and supplementary lighting; this trend is thought likely to continue and even increase in a new generation of smaller but more lavishly equipped cars.

Since much of the discussion of the recyclability of vehicles takes place in the context of copper, it is appropriate to examine ways in which this problem might be eliminated through engineering design. In many cases, this is inextricably bound up with the question of materials substitution. Copper is not the only problem, but its examination will illustrate the complexity of the exercise.

The problem of copper contamination may be resolved through the following:

(a) materials design, to replace copper by harmless materials, or
(b) mechanical design, to permit easy removal of copper.

5.11.2.1 *Materials design*

One method of reducing copper contamination is to replace it by a non-contaminating material. From the data of Dean and Sterner, it is evident that copper was used in the composite car principally as:

(a) Heat exchangers (7·0 kg)
(b) Electrical wiring (6·4 kg)

5.11.2.1.1. *Heat exchangers*

Copper and brass are used in heat exchangers primarily for their good thermal conductivity but also for their corrosion resistance and ease of joining. Typically, traditional copper vehicle radiators contain about 70% Cu, with the balance comprising lead, tin and zinc, all of which are unwelcome contaminants in steel and in steelmaking. Radiators have continually been developed to reduce the finished cost, usually by gauge reductions, and further materials savings result from the use of copper foil in fins. They are, however, assembled by labour-intensive means and, although work continues to develop one-shot soldering, manufacturers have sought other materials and methods of construction.

However, largely for financial reasons at a time when the price of copper rose — temporarily, as it emerged — to more than twice that of aluminium, there has been a major move towards aluminium-based radiators. By the late 1960s the price differential was sufficient to justify the replacement of copper by aluminium radiators. They now arise sufficiently often for them to be specifically excluded from radiator scrap specifications.

The essential requirements for a radiator material are:

(a) thermal conductivity,
(b) mechanical strength,
(c) ease of joining, and
(d) corrosion resistance.

The materials normally considered for this application are aluminium and Al 1·25% Mg, which offers improved strength. Their thermal conductivity is only 50% that of copper but is 30% better than that of brass.

Tensile strength is 20–75% that of brass and copper but, since specific gravity is some 33% of that of copper or brass, even with the necessary increases in thickness, weight savings of up to 30% can be achieved.[58]

The joining of aluminium alloys is complicated by the presence of the tenacious oxide skin, which inhibits wetting by brazes or solders. Fluxed brazing was used for radiators in the 1950s but was abandoned because of the labour costs associated with efficient removal of the corrosive flux.[59] Fluxless brazing under vacuum requires integral cladding, at the hot-rolling stage, with an Al 7·5–11% Si 1–2% Mg alloy to a thickness of 5–15% of the core thickness.

Mechanically jointed radiators are in use, notably by Volkswagen. Such radiators are of complex construction, and in addition to aluminium, copper or brass tubes, may contain steel end plates and plastic tanks.

Adhesive bonding of radiators offers the advantages of low process temperature and the avoidance of an additional metal/metal junction as a site for corrosion. Doubts remain concerning the likely service life of the adhesive and of the thermal insulation effects of the resin films at the joints.

All-aluminium construction is well-established for oil coolers and air conditioner units, where only an external corrosion hazard exists, as compared with radiators and heaters, where solders and brazes can set up internal galvanic corrosion. Inhibited antifreezes have protected aluminium in a mixed metal system during limited field trials.

A general reluctance to change materials and methods of construction may stem from doubts concerning long-term service life. The substantial capital still invested in copper-brass production equipment will, it is thought, ensure the continuance of the traditional type for some time, albeit in reduced numbers.

The value and relative ease of removal of the radiator, together with a steady market for such copper/brass radiator scrap, virtually assure that it will be stripped from hulks. Of all the copper-bearing components in the car it carries the least risk of being incorporated in a bale, although oversight or entrapment in a damaged vehicle could increase residual copper content by an additional 0·5% or more, allowing for the fact that the weight of the residual steel is very much less than that contained in the original vehicle. The avoidance of such uncertainty would be an added advantage of the aluminium radiator.

5.11.2.1.2 *Electrical uses*

Copper wiring may be replaced by aluminium, which has, however, only about 61% of the electrical conductivity of copper and which, for equal current-carrying ability, must have 1·64 × the cross-section of a

[58] J.E. Tomlinson, Progress in aluminium automotive heat exchangers, *Sh. Metal Inds.*, **51**, (12), pp.766–768, Dec. 1974.
[59] Alan Baker, The aluminium radiator, *J. Automot. Engng.*, **6**, (2), pp.33–5, Apr. 1975.

copper conductor. With its low specific gravity, 1 kg Al could replace about 1.5 kg Cu as a conductor; typically, about 2·5 kg of electrical and motor wiring could be replaced by aluminium.

Increased wire section requires larger conduits, larger holes, and greater clearances behind instrument panels to facilitate bending through necessarily larger radii. Aluminium is more difficult to bend than is copper wire. A further complication has been the difficulty of making adequate permanent junctions between aluminium wires. The tendency of aluminium to oxidise means that the wires must be thoroughly cleaned before being clamped together. The gradual reformation of the skin, with a consequent increase in resistance, produces localised heating; so great is this risk that domestic aluminium wiring was for a time prohibited in the U.S.A. However, the use of mechanical clamping devices is well advanced, with provision of features designed to disrupt the oxide film and to prevent flattening of the soft aluminium wire during the clamping process. The coating of copper terminals with tin prior to fusing them to the aluminium wire is effective in reducing corrosion at copper/aluminium junctions, but additional encapsulation with plastics, silicones or waxes is desirable. Such preconditioning of wires is not intrinsically difficult.[60]

Further problems have arisen with the low creep strength of aluminium under the centrifugal forces of service as a motor winding, possibly at temperatures as high as 100°C; additions of 0·75% Fe and 0·15% Mg effect a marked improvement.[61]

Aluminium was used in wiring harnesses as early as 1951;[62] the main factor preventing its widespread use in this area is cost, since neither the light PVC-insulated cables nor the enamelled aluminium already used in starter motors and alternators is cheaper than copper. The costs of the heavier battery leads are more favourable to aluminium, but fatigue life is unsatisfactory unless the wires are stranded. It has, however, long been used in the paper-insulated windings of the stators in heavy-duty starter motors.[63]

Many of the major car manufacturers and the aluminium companies have investigated the electrical use of this metal. In the mid-seventies, of the General Motors models built in the U.S.A. only the Corvette did not have an aluminium front wiring harness.[64]

It must be concluded that existing knowledge would readily permit the use of aluminium wiring in vehicles.

[60] Allan Warner, Tang termination of aluminum wire, *Proc.1972 Fifth Annual Connector Symp.*, Cherry Hill, N.J., Oct. 18–19 1972.
[61] Arthur G. Craig, What you should know when using aluminum magnet wire for motor coils, *Insulation/Circuits*, pp.38–41, Feb. 1973.
[62] J.S. Poliskin, Automobile steel scrap of low-residual copper, *Proc.27th Electric Furnace Conf.*, *Metallurgical Soc., Am.Inst. of Min., Met. and Pet. Eng., Detroit*, pp.141–2, Dec. 10–12, 1969.
[63] Aluminium Wire and Cable Company Ltd., private communication.
[64] Harry A. Stark, (Editor), *Ward's 1976 Automotive Yearbook*, 38th Edition, p.65, Detroit, Ward's Communications, Inc., 1976.

5.11.2.1.3 *Permanent magnets*

A further reduction in copper usage in vehicles is possible through the use of permanent magnets to replace copper-wound electric motor stators. Advances in magnet technology since World War II have expanded the range of available materials, and ferrites, with their high coercive forces, are well-suited to use in small electric motors. They are, without exception, very hard and brittle and are therefore likely to be ground to powder during shredding and not to enter the steel. Barium and strontium ferrites are well-known examples of the type, and are widely used in stators.[65,66]

Alternative materials might include the Alni group (Fe–Al–Ni or Fe–Al–Ni–Cu), or the Alnico (Fe–Al–Ni–Co) group.

5.11.2.2 *Engineering design*

Certain copper-bearing components are inaccessible and difficult to detach, and design changes to improve the situation would provide easier and cheaper maintenance. However, since the overriding concern of the manufacturer is cost, existing designs are likely to be the most cost-effective ones in manufacture.

Possible design changes to facilitate copper removal have been suggested by Stone, who considered the car as three sections:

(a) under-bonnet area,
(b) under-dashboard area, and
(c) body, boot and tail-light areas.

It is throughout assumed that financially attractive components such as the engine, radiator, and dynamo will already have been removed before the hulk arrives at the scrapyard.

5.11.2.2.1 *Under-bonnet area*

Although most components and wiring are visually and manually accessible, especially on removal of the engine, they lack the non-ferrous scrap value to offset additional handling costs. The problem may be solved by one of two methods:

(a) clustering the copper-bearing components onto larger or more valuable components now routinely removed, i.e engine and radiator, or

(b) attaching almost all copper-bearing components to one or two easily removable brackets or mounting plates.

One mounting plate could be near the inside front of one wing near the headlight, and could be a common location for the starter solenoid, horn, voltage-regulator and relays for high-current accessories such as air conditioners. Such proximity to the battery implies shorter runs of heavy-

---

[65] J.S. Poliskin, *loc. cit.*
[66] H. Zijlstra, Permanent magnets, *Physics in Technology*, **7**, pp.98–107, May 1976.

gauge wire. Attachment of the plate to the radiator would ensure its removal with the latter. A mounting plate on the firewall could carry the heater core and motor, windscreen-wiper motor, ignition coil and other relays.

Consolidation of electrical components in two places would, in turn, confine the major wiring to one side of the vehicle, with additional minor wiring for the light and direction indicator cluster at the other side. The wire loom and secondary wires could easily be removed, and with the ignition system normally removed with the engine, would essentially eliminate all wires beneath the bonnet.

Under-bonnet components are scattered about the engine compartment but their removal, even without re-design, requires only a few moments with cutting torch or crowbar. Relocation and consolidation would, however, encourage all yards to remove them, simply because it would be worthwhile to do so and, in practice, almost impossible to avoid so doing during the process of removing the radiator and heater core.

5.11.2.2.2 *Under-dashboard area*

Wiring in this area is associated principally with inaccessible instruments, radio and fuses; it is generally unprofitable to recover the metal. Redesign could include:

(a) provision of easy access,
(b) a common mounting plate for instrumentation and accessories,
(c) standardisation of the fuse-box location,
(d) consolidation of wiring,
(e) push-type terminal connections, or
(f) possible replacement of wiring by a printed circuit board for applications drawing only light currents.

Removal of this wiring still leaves the copper contained in the instruments, of which the mounting and siting varies widely from one vehicle to another at the whim of the stylist. One solution which would allow freedom of dashboard styling is the development of a circuit-board type of attachment carrying all the instruments; thus, the whole assembly could be removed as a unit.

A group of light-gauge wires passes within the steering column and controls equipment such as direction indicators, horn, lights and possibly windscreen wipers and windscreen washers. Stone has suggested methods for their recovery, but none seems justifiable in view of the very small amount of copper involved.

5.11.2.2.3 *Body wiring*

Generally, the solution to this problem is to provide better access and easier detachment for accessory motors and wiring; such refinements would also aid in maintenance.

Many accessory electric motors are difficult to find and remove. Each window motor can contribute more than 0·2 kg Cu, and each seat motor, mounted on the seat or on the vehicle floor, contains up to 0·3 kg Cu. Since seats are almost always removed in dismantling, standardised mounting of the motors on the seats would help the recovery process as a whole.

Remaining body wiring and that of the tail-light section could be consolidated in a plastic conduit mounted beneath the car and connected by pull-away type clamps for simple removal. Another possible method of wiring consolidation is the use of a common flat wire, with break-away type connections on each branch wire.

It should not be overlooked that any method for easy removal of wiring is likely to increase the scope for malicious damage of the car, since it would be very easy for vandals or children to immobilise it.

5.11.2.2.4 *Other suggested redesign considerations*

(a) A change from a 12 to a 24 V system would result in cost reductions in a reduced content of copper in wire and in accessory motors. The 24 V system is already used in aircraft and in some heavy commercial vehicles. It requires a 24 V battery which would imply a unit of approaching twice the size, weight, and cost of the current 12 V units. The cost savings in copper would not offset additional battery costs.

(b) Light transmission through fibre optics; could eliminate copper in information-transmitting circuits.

(c) Location of front and rear lights within or attached to front and rear bumpers would allow wires supplying them to be located beneath the chassis, where they would be accessible. Bumpers are normally removed prior to hulk processing; many are even straightened, where necessary, replated, and sold as replacement parts. The lights would, however, be more vulnerable to collision damage.

(d) The replacement of electric motors by pneumatic or hydraulic controls is technically feasible. This would obviate the location of copper wires in doors and other inaccessible areas, and would also eliminate the window, seat, wiper and aerial motors. A central compressor would be required, which would itself need to be non-contaminating.

(e) It has been suggested that the increasing use of plastics in, for example, dashboards, would allow copper wires to be recovered after incineration. This procedure would require gas-cleaning equipment.

(f) There is already a strong move — which will almost certainly continue — towards miniaturised and printed circuits, which reduce the copper content. Their use in many applications, from calculators to audio components, has demonstrated an ability to reduce maintenance costs by virtue of their robustness and ease of replacement.

5.12   THE CATALYTIC CONVERTER

The catalytic converter used in vehicles to reduce the amounts of pollutants emitted by road vehicles is a major consumer of valuable metals. As a valuable and identifiable component, it is unlikely to be discarded knowingly.

Estimates in the U.S.A. suggest that 5–7000 tonnes of chromium per annum might be recoverable from catalytic converters once their use becomes universal in passenger vehicles. It is also reported that about 115 000 troy oz. of platinum was recovered from such converters in 1982, with estimates of 500 000 oz. by 1995.[67]

It should be noted that the automotive catalytic converter consumes a very large share, 37%, of total platinum production and an even larger one, 68%, of rhodium production. This subject is developed in Chapter Nine.

5.13   DISPOSAL OF RESIDUALS

Any material that is not reclaimable from the vehicle is a residual of recovery, and must be disposed of in some way. The disposal costs must be considered as part of the overall economics of the recovery operation. Residuals will arise in any of the recycling operations discussed in this work. However, given the relative importance of the automotive sector, it is appropriate to discuss residuals disposal here.

The residuals of vehicle dismantling and shredding operations are largely non-metallic *detritus*, which tends to be of low density and hence expensive to transport and landfill. The 1960 composite car contained only 0·9 wt.% polymers; in the model years 1972 and 1973, U.S.-built cars typically contained almost 5% polymers, largely used in safety and comfort applications. Since the first oil crisis, attempts to lighten cars have still further increased the extent of replacement of metals by polymers, with a corresponding increase in the quantities of non-metallic waste at the end of the life of the vehicle. It is highly desirable that means should be found of alleviating the financial burdens associated with this residue.

As seen earlier, increased quantities of non-metallic residuals of the vehicle processing operation and, therefore, increased disposal costs, are accompanied by reduced revenues as metals are replaced by substitutes which currently have no scrap value. The economic pattern of hulk processing is, therefore, moving towards instability.

The following four general methods have been advanced for commercial utilisation of the light fraction.

[67] LaVerne Leonard, Specifying metals for recycling, *ME,* pp.47–50, Sept. 1985.

5.13.1 Landfill

It is unlikely that any revenue could be generated by this means, since the non-metallic residue is voluminous and compressible. Therefore, it lacks the properties desirable in hard-core. Transport costs and dumping fees would be incurred in disposal of the material by this means.

5.13.2 Secondary materials: polymer-enriched fraction

The potential problems likely to arise from increased incidence of valueless plastic wastes caused concern as long ago as 1974. At that time, it was considered economically feasible to hand-dismantle to recover polyurethane foam for recycling.[68] Such foams are particularly intractable, on account of their low bulk density of $33 \cdot 6$ kg/m^3. Attempts have been made to separate them from shredder residues, by flotation. The results showed that the recovered foam could be hydrolysed to reusable liquid mixtures of polyether glycol monomers and toluene diamine.[69] [70] Both Ford[71] and General Motors[72] have also published details of hydrolysis processes for such foams. Polyurethane waste foams are the subject of a Japanese patent, which describes a process to convert them into a foam moulding which gives good elasticity.[73] vxed plasticised, rigid and elastomeric polyurethane foams have been processed into polyurethane, epoxy coatings and urethane or epoxy polymer concrete.[74] However, so far as is known, no commercial process has yet gone into operation, presumably because of unattractive economics.

Clean polymeric scrap is a valuable material that is routinely recycled. Methods have been devised for cleaning and regranulating even contaminated wastes such as agricultural film. However, it is recognised that, with current technology, it is not financially viable to separate individual plastics from mixed wastes, such as municipal solid waste (MSW) or from scrap shredder residues, to provide fractions which can be reassimilated into the materials stream for applications similar to those of the *original*, as opposed to a degraded, use. It has been estimated that in the period 1985–1990 only about 25% of the total quantity of plastics waste can be diverted from the municipal waste stream. Of this some 27%, i.e. approximately 7% of the total, is likely to come from the transport sector in the U.S.A.[75] Numerous possible methods exist for reutilisation of the poly-

[68] K.C. Dean, J.W. Sterner and E.G. Valdez, *Effect of increasing plastics content on recycling of automobiles*, Washington, D.C., U.S.Department of the Interior, TPR 79, May 1974.

[69] E.G. Valdez, K.C. Dean and J.H. Bilbrey, Jr., *Recovering polyurethane foam and other plastics from auto-shredder reject*, Washington, D.C., U.S.Dept. of the Interior, Bureau of Mines, RI 8091, 1975.

[70] E.G. Valdez, Separation of plastics from automobile scrap, *Proc. Fifth Mineral Waste Utilization Symp.*, Chicago, pp.386–392, Apr. 13–14 1976.

[71] Lee R. Mahoney, Steven A. Weiner and Fred C. Ferris, Hydrolysis of polyurethane foam waste, *Environmental Science and Technology*, **8**, (2), pp.135–139, Feb. 1974.

[72] Carol K. Lewicke, Getting the most out of polymer scrap, *Projects*, General Motors Research Laboratories, (PR-236), Apr. 1974.

[73] Sadao Kumasaka, Recovered scrap-containing polyurethane foam, *Jpn.Kokai Tokkyo Koho* JP 61,272,249 [86,272,249] (Cl.C08J9/22). Dec. 2 1986.

[74] F. Vohwinkel, *Farbe Lack*, **93** (1), pp.10–13, 1987.

[75] T. Randall Curlee, The Recycle of Plastics from Auto Shredder Residue: Incentives and Barriers, *Materials and Society*, Vol.9, No.1, pp.29–43, 1985.

meric materials in the automobile.[76, 77] Fibreglass-reinforced polyester composite wastes can, in principle, be treated to recover the glass fibres after hydrolysis of the resin. However, the recovered fibres cannot be recycled into resin composites without first applying a surface treatment.[78]

5.13.3 Secondary materials: wood substitute and other structural materials

Mixed plastics scrap has been converted, notably in Europe and Japan, to a variety of useful products including fencing, pallets, piling, and moulded items. Various pieces of equipment have been designed specifically to process complex mixtures of plastic scrap, but this has usually been in-plant scrap.[79]

The use of post-consumer scrap is a much more intractable problem. There is much effort worldwide to develop methods for the processing of heterogeneous wastes from plastics and fibrous materials, from metal-plastic composites, from laminates, and from residues such as waste tyres. One approach to the problem has been made by the Plastics Institute of America, with support from the Department of Energy, to convert polymers recovered from shredder residues into structural panels. Dirt, glass, residual metals and stones must first be removed. The panels exhibited properties comparable to those of commercial masonite or particle board and better than those of ceiling panel; they were, however, inferior to those of plywood. The small amount of thermoplastic material present sufficed to bond the material, but only when adequately dispersed throughout the mass. Hence, reduction in particle size, e.g. by pulverising, is essential to facilitate both compression and injection moulding tests.[80]

The results of the tests showed that plastics scrap can be fabricated into useful structural shapes, but that it is desirable to improve their properties by the use of bonding agents, by blending with a scrap material with an increased content of thermoplastics, or by blending in small amounts of inexpensive resins, such as PVC. Various proprietary binders have been evaluated in the preparation of experimental building panels. Properties are claimed to be superior to those of particle board or masonite.[81]

[76] Maximilian J. Wutz, Plastic Material Recycling as Part of Scrap Vehicle Utilization - Possibilities and Problems, *Conservation and Recycling,* Vol.10, No.2/3, pp.177–184, 1987.

[77] Kiyoaki Kumano, Reutilization of Motor Vehicle Molding Wastes, *Clean Japan,* No.12, pp.1–5, Feb. 1988.

[78] J.M. Bouvier, S. Esperou du Tremblay and M. Gelus, Managing Fiberglass-Reinforced Polyester Composite Wastes, *Resource Recovery and Conservation,* Vol.15, No.4, pp.299–308, Nov. 1987.

[79] See, for example, *Plastics Recycling '88,* Society of Plastics Engineers Scandinavia Section and Department of Polymeric Materials, Chalmers University of Technology, Copenhagen, May 17–19 1988.

[80] Plastics Institute of America, *Research Report, Shredder Residue,* Dec. 1983.

[81] Rudolph D. Deanin and Chaitanya S. Nadkarni, Recycling of the mixed plastics fraction from junked autos, I.Low-pressure moulding, *Proc. Meeting of American Chemical Society, Polymeric Materials Science and Engineering Division,* St.Louis, Apr. 11 1984.

5.13.4 Incineration

Non-metallic shredder residues have a worthwhile heat value, estimated at 16·6 MJ/kg, compared with about 30 MJ/kg for anthracite.[82] Other studies suggest an even higher heat value, of 24 MJ/kg.[83] This could, in principle, realise residual energy which could be used to generate in-plant steam, or even electricity, and so improve the economics of the scrapyard. The disposal of such residues may thus become a process of exploitation, if performed in a suitable manner.

The thermal degradation of plastics, either by hydrolysis to chemicals or by incineration for energy recovery, is open to the same criticisms as that of any other structural material, in that one should, if possible, retain them in their most useful — that is, their most highly ordered — state for as long as possible through as many uses as possible; in this case it would be as structural materials. If, however, they are too badly contaminated for recovery and recycling in structures, thermal degradation is arguably better than landfill.

However, despite their worthwhile heat values, shredder residues cannot currently be exploited economically because:

(a) The material is not readily combustible. The work of Daborn and Webb showed it to have, as screened and air-classified, a very high (50%) ash content. Though capable of burning, it has been found to generate large quantities of slag which, when solidified, was difficult to extract.[84]

(b) The material contains certain components, of which PVC has given particular cause for concern, which release corrosive and, of perhaps greater concern, polluting fumes. Some 80–90% of these can be removed by suitable techniques, but the need for scrubbers may so increase the cost as to cancel any financial gain from incineration, certainly at a scrapyard level. Some recent literature sources on the problem are examined below.

In a general examination of hydrochloric acid emissions attributable to the incineration of PVC wastes in Western Europe, Lightowlers and Cape identify the sources of chlorine compounds in municipal and other incinerators as PVC, vegetable matter, dry cell batteries, and sodium chloride. Natural sources, such as volcanos, also produce the gas.[85]

There is some divergence of opinion on the extent to which PVC contributes to emission of HCl from incinerators. Buekens and Schoeters point out that although the addition of PVC to refuse has increased HCl

[82] K.E. Boeger and N.R. Braton, Mill fuel and mill cover recycled products from shredder fluff, *Resources and Conservation*, **14**, p.133, Mar. 1987.

[83] G.R. Daborn and M. Webb, *Treatment of fragmentizer waste by starved air incineration — a brief feasibility study*, Department of Industry, Warren Spring Laboratory, LR 465 (MR) M, Oct. 1983.

[84] *ibid.*

[85] P.J. Lightowlers and J.N. Cape, *Hydrochloric acid emissions attributable to the incineration of PVC wastes in Western Europe*, Inst. of Terrestrial Ecology, APME/NERC Contract, ITE Project 1030, Final report to APME, Apr. 1986.

emissions, so also have additions of sodium chloride. The fraction of the HCl emission due to PVC is impossible to distinguish from that from other sources. There are indications that HCl emissions can take place in the complete absence of PVC.[86]

There is at least some controversy, also, over whether PVC can, in thermal degradation, serve as a precursor of polychlorinated dioxins (PCDD) and dibenzofurans (PCDF).[87]

HCl emissions from incinerators are very variable. PVC is a minor constituent of both municipal solid waste and of shredder residues. It contains about 58% wt.% Cl but since at least half of the PVC in UK domestic waste is likely to be plasticised, and so contains some 30% plasticiser, due allowance must be made for PVC dilution. About half the chloride content of waste in municipal incinerators may arise from sources other than PVC.

Calculations for the U.K., where no municipal incinerators are known to have flue-gas washing facilities, suggest that PVC waste incineration — of all kinds, not specifically of shredder residues — currently accounts for 6% of total HCl emissions and for 0·1% of the *total* potential acidity. This level would be expected to rise if shredder residues, few of which are presently incinerated, were to be burned in incinerators *not equipped with gas-scrubbing facilities*. It should be noted that the U.K. levels noted above are low in the European context. Corresponding figures are:

Austria (28%/0·1%); Belgium (29%/0·3%); Denmark (59%/0·4%); Eire (0%/1·1%); France (54%/1·7%); Italy (41%/0·1%), Netherlands (53%/0·6%); Sweden (75%/0·7%), Switzerland (94%/2·3%) and West Germany (27%/2·2%).

All produce more of their total acidity from PVC than does the U.K. According to these calculations, only Spain (5%/0·02%) and Eire have a lower figure than the U.K. for HCl emissions attributable to PVC.[88]

The scope of the present work does not permit a full analysis of the nature and magnitude of environmental pollution which might result from the incineration of PVC. What is indisputable, however, is that PVC is one of several constituents of shredder residues which could yield combustion products whose release into the atmosphere would be considered undesirable or unacceptable. As such, its elimination from them could only be beneficial in any attempt to devise an incineration process for residues at a scrapyard level.

[86] A. Buekens and J. Schoeters, Refuse incineration and PVC, *Proc. European Conference on Plastic Packaging,* Brussels, June 2/3 1986.

[87] Christoffer Rappe, Polychlorinated dioxins (PCDDs) and dibenzofurans (PCDFs). Occurrence, environmental levels and formation in thermal processes, *Proc. European Conference on Plastic Packaging,* Brussels, June 2/3 1986.

[88] P.J. Lightowlers and J.N. Cape, *op. cit.*

5.13.5 Conversion to fuel

In many countries throughout the world, there is considerable activity into the conversion of waste plastics and rubbers, by thermal cracking, into light and heavy fuel oils.

So far as is known, none of the expedients of 5.13 has yet shown promise of technological and potential profitability. At present, it would appear that the best that may be hoped for is the development of some conversion/reutilisation process that could reduce the overall costs of disposing of automotive residues.

6
RECOVERY OF MATERIALS FROM APPLIANCE SCRAP

6.1 INTRODUCTION

From an environmental viewpoint the disposal of domestic appliances ('white goods') poses an obvious and conspicuous problem. Unlike unserviceable cars, which are difficult to transport for illicit dumping and are, in any event, registered by the authorities, appliances are relatively portable and can be disposed of in woods, ditches, or rubbish tips.

The feed to scrap processing yards has always included a proportion of material from domestic appliances. For reasons to be discussed, they have not always been desired and numerous processors have declined to handle them. However, when other forms of scrap, notably car hulks, have become scarcer in times of high industrial activity, appliances have formed an important fraction of the input to shredders and balers and have been processed along with the other material included, especially automotive scrap. By its complex nature, appliance scrap suffers from many of the disadvantages of automotive scrap. However, it is increasing in its contribution to overall scrap flow. It has even been suggested that it has a beneficial effect when blended with automotive scrap, in evening out power requirements in the shredder, so reducing electrical energy costs.

6.2 RECOVERABLE MATERIALS IN APPLIANCES

There is no shortage of data on the statistics of appliance numbers in the U.S. and Europe, though recent data are sparse.[1,2] Future quantities of discarded appliances for potential recycling depend on market saturation, technological innovation, and on the life expectancy of the appliances. All of these are the subject of periodic assessment and projection.

Estimates of the overall tonnage of domestic appliances available for recycling in 1980 in the EEC suggest some 1·7 million tonnes, of which 1·4 million tonnes was estimated to be ferrous metals.

By contrast, there are few data on appliance composition. The most recent traceable analysis of composition dates from 1971 and relates to

---

[1] See, for example, Europool, *The Disposal and Recycling of Scrap Metal from Cars and Large Domestic Appliances,* London, Graham & Trotman Ltd., 1978.
[2] Edwin A. Kinne, Options for the collection and recovery of household appliance materials, *Proc. Sixth Mineral Waste Utilization Symp., Chicago,* May 2–3 1978.

Table 6.1
Materials use in selected appliances and value at 1976 scrap prices[4]

Appliance	Materials (kg) (average)						Value ($)
	Steel	Cu & alloys	Al & alloys	Glass	Polymer	Paper	
Room air conditioner	28·1	16·3	4·5	–	4·1	3·6	15·01
Kitchen range*	80·8	0·9	0·9	5·4	0·9	0·9	5·52
Refrig.	117·9	5·4	4·1	4·5	15·4		11·86
Dishwater	54·4	2·3	0·9	–	9·0		4·98
Washer**	93·9	1·8	6·8	0·9	3·2		8·87
Drier	59·9	0·9	1·8	0·4	2·7		4·59

* Also contains 0·9 kg zinc
** Also contains 6·8 kg concrete

conditions in the U.S.A.[3] At that time, major appliances typically weighed around 90 kg, with a density of 160 kg/m^3. Materials use in appliances is given in Table 6.1.

The total monetary value represents that of the materials after removal and segregation. The high value of air-conditioners results from their high copper content. Driers, at $4.59 have the lowest value.

Generally, scrap values per unit in appliances would be low even if removal and separation of valuables were easy. In fact, the variety of materials used and the improved manufacturing methods that have steadily reduced real prices of appliances have simultaneously made them more complex and have reduced the quantity and value of metals in them. Only processors employing the most advanced technology can reclaim materials economically, and this is reflected in the low prices offered for appliances and, until recently, the lack of interest shown in them by scrap processors.

6.3 CURRENT RECOVERY PRACTICE

Appliances tend to suffer the same depredations as cars in respect of high value components. They are cannibalised for serviceable electric motors, burners, compressors etc., and for obvious pieces of non-ferrous metal. They are then often left to rot in the open air. In scrapyards, they are routinely processed alongside cars.

---

[3] Anon, *The disposal of major appliances*, Report prepared by the National Industrial Pollution Control Council for the Secretary of Commerce, Washington, D.C., U.S.Government Printing Office, June 1971.
[4] Anon, *The disposal of major appliances*, Report prepared by the National Industrial Pollution Control Council for the Secretary of Commerce, Washington, D.C., U.S.Government Printing Office, June 1971.

6.4 THE NATURE OF APPLIANCE SCRAP

Appliance scrap is processed with cars and light gauge metal, with which it it shares a number of characteristics.

6.4.1 Steel from appliance scrap

The same problems of contamination may be expected as those encountered with road vehicles, derived from the variety of materials used and their incompatibility in the reclaimed steel. In addition, appliances commonly carry additional contamination in the form of porcelain enamel.

Steel covered with porcelain enamel as contaminant in; has traditionally been excluded from any form of commercial scrap. Even the specification for the No. 2 bundle, a grade known to be highly contaminated, carries a note 'no vitreous enamelled material'.[5] For many years, household appliances, which form an arising known as 'white goods', were virtually ignored by scrap merchants.

The reason for the specific exclusion of such goods lies in the enamel coating, which may represent 5 wt.% or more of the scrap. The coating consists mainly of silica, with mixtures of compounds of elements such as antimony, cobalt, lead, titanium and zinc, most of which are undesirable in steel. However, the base metal is almost invariably steel of deep-drawing quality. In the absence of the enamel, the steel could be processed by a scrapyard to achieve a high grade.

Starting in 1972, the Whirlpool Corporation, a major American producer of domestic appliances, began an exploratory programme with Inland Steel to examine the possibilities of using porcelain-enamelled scrap.[6] A special collection of scrapped home appliances was made by Whirlpool and was delivered to Inland towards the end of 1974 to evaluate its suitability for steelmaking. The scrap was unprepared except for pressing it into bundles; no copper motors, windings or polymers were removed prior to pressing.[7]

Owing to the uncertain properties of the scrap and to its expected level of residual copper it was regarded as a substitute for No. 2 bundled scrap. Ten open-hearth heats were made to different hot metal levels, replacing all or almost all of the usual No. 2 bundled scrap with a total of 360 tonnes of appliance scrap. No problems developed with furnace refractories, productivity, flux consumption or residual sulphur or phosphorus. At levels of 8–12% of total metallic charge, copper-bearing steels containing 0.20–0·30% Cu were produced. It was accepted that the market for such

[5] *Specifications for iron and steel scrap 1975*, Washington, D.C., Inst. of Scrap Iron and Steel, Inc., 1975.

[6] The stimulus for this programme was probably a shortage of other forms of scrap and a realisation that discarded appliances were beginning to pose environmental problems.

[7] Inland Steel, Use of porcelainized scrap in steelmaking, *Porcelain Enamel Inst. Forum*, Nov. 4, 1975.

[8] T. H. Goodgame and E. W. Hartung, Progress in resource recovery in appliance manufacturing, *Proc. Sixth Mineral Waste Utilization Symp.*, Chicago, p.411–418, May 2–3 1978.

steels was limited and, therefore, that unprepared appliance-derived steel is of limited utility.

The results of the study must be regarded as inconclusive, since the proportion of non-metallics in the scrap was not known, and it is admitted that the programme was only a limited one. However, there was no evidence of unacceptable levels of antimony, boron, chromium, cobalt, nickel, niobium, phosphorus, sulphur, tin or zinc. At this modest level of scrap use the steels met the requirements for cold-rolled commercial quality (CQ) steel. So far as deep-drawing qualities are concerned, the limiting factor is almost certain to be copper.

Since the trials, which continued until at least 1976, reported by Inland, appliance scrap has become a routine source of feed to scrap shredders. Initially, it was included for lack of more traditional sources, such as motor vehicles. It is probable that much of the enamel is broken away from the steel surface during shredding, and does not therefore enter the melt in the quantities that would be involved had the scrap been prepared simply by pressing. In some scrapyards, white goods are estimated to comprise some 20% of total input tonnage; however, since few if any shredder operators have any particular need to note such statistics, this should be regarded more as an impression, if not a guess.

Whirlpool has also sponsored an experimental dismantling programme, to determine times of dismantling and the values of recovered materials.[8]

6.5 CHANGES IN MATERIALS USE IN APPLIANCES

The greatest single change in the domestic appliance market over recent years has been the substitution of metallic by non-metallic materials, in particular by the so-called NIMS (New Inorganic Materials Science). This is a microdefect-free material developed by ICI. It uses sand as a filler.

Use of this material will permit considerable financial savings in manufacture and, eventually, to the consumer. For example, a hob made in metal might retail at £50, having cost £20 or less to produce. The use of NIMS might reduce manufacturing cost to £8, a point at which the operation is nearing the throw-away stage when repair is necessary.

In a cooker, outer panel temperatures never exceed 120°C, and sheet metals can therefore be replaced by NIMS. A new gas cooker weighing perhaps 50 kg may currently contain 40 kg steel and 10 kg controls. With NIMS, this may reach 20/20, with a little aluminium, brass and plastic (in the case of a cooker without a glass panel.) Higher-quality cookers weighing perhaps 60 kg may in future contain 20 steel, 15 NIMS, 10 glass, 10 controls, and miscellaneous.

By law, front panel temperatures must not exceed 80°C, so polycarbonate is a possible material. It will not, though, withstand scratching and the effect of kitchen abrasive materials.

It is thought that all post-1989 appliances made by the technology licensee will feature the substantial use of NIMS. Production lifetime of many appliances is already down to about 18 months, emphasising the rate of technological change.

Metal gauges in appliances are steadily diminishing. Typically, 0.6mm gauge steel is used. The philosophy is to reduce the cost of, for example, a hob to less than the cost of a £17.50 service call. The use of quick-release, self-sealing fittings will then permit the use of totally unskilled service personnel. One large company utilising it envisages that NIMS will replace all steel sheet in its appliances by about 1995. Gears are already largely made of non-metallic materials such as nylon.

6.6   BARRIERS TO APPLIANCE RECYCLING

The barriers to appliance recycling are essentially the same as those to automotive recycling. Appliances contain a wide and self-contaminating variety of materials. Further, as in road vehicles, units are becoming lighter and materials which can be recovered and sold are being replaced with substitutes, whose ultimate disposal is at the cost of the reclaimer. Inasmuch as safety considerations do not require minimum strength, as would be required for road vehicles, one may expect steel to disappear from appliances more rapidly than from cars.

There is also the problem of collection, which must be solved if economies of scale are to be achieved in processing appliances.

6.7   POSSIBLE DESIGN CHANGES TO FACILITATE RECYCLING

6.7.1   Engineering design

Increasingly, appliances are being controlled not by complex and potentially unreliable electro-mechanical switches and timers, but by microprocessors as sources of contamination. As pointed out earlier, these contain small quantities of materials whose use is relatively recent, and whose effects as tramp elements in the metals with which they might be co-recovered are unknown. Since, in any event, such materials are of high unit value, it is doubly desirable that means should be provided for their easy identification and removal. Consideration should be given to the enclosure of microprocessors in a conspicuously identified case, e.g. one marked with fluorescent paint, and to making connections to such microprocessors by means of easily removable connectors.

6.7.2   Materials design

As suggested earlier, changes in materials are most unlikely to be made for the sake of recyclability. If the replacement of steel by NIMS, or similar non-metallic materials proceeds as expected, the recyclable value of major appliances may be expected to decline very rapidly. This might, however, actually be an advantage inasmuch that recovery from appliances will then be a question of removal of valuable components or of obvious pieces of non-ferrous metals, followed by the discard of a residue that contains essentially no valuable materials. If, by oversight, an electric motor or microprocessor were left in the appliance, there would then be no possibility of its entering into a subsequent steelmaking heat.

7
RECOVERY OF MATERIALS FROM ELECTRONICS SCRAP

7.1 INTRODUCTION

Electronic scrap is one of the most important sources of raw material for the precious metal refiner. Electronic equipment may contain, in aggregate, significant quantities of precious metals and other recoverable materials. On sample batches of 36 separate electronic components available for scrap recovery from defence sources, it was concluded that the recoverable precious metals gave the scrap an average value of $0.22/lb, at 1978 prices.[1] However, the nature of electronic goods is changing rapidly, and such changes are likely to have significant impact on future recovery operations.

7.2 RECOVERABLE MATERIALS IN ELECTRONIC GOODS

After a life of at least 25 years, much of the existing telecommunications equipment is no longer cost-effective. The specifications for such equipment called for 200 microinches of gold. Telephone contacts, of which each instrument has many, each contain 2–10 cents worth of gold.

Mixed telecommunications scrap typically contains less than 1 troy oz. of gold, 3–10 troy oz. of silver, and 1–7 troy oz. of palladium per tonne of scrap. Segregated, higher-grade telecommunications scrap, such as plugs, may contain 200–600 troy oz. of gold/tonne. Mainframe computers may contain 10–20 troy oz. of gold, approximately 100 troy oz. of silver, and several tons of copper, together with some steel and quantities of valueless polymers and paper.

7.3 SOURCES OF ELECTRONICS SCRAP

As indicated above, consumer electronics scrap generally exists in small units whose distribution is widespread. Such objects hardly contribute in any practical way to the activities of the precious metals reclaimers for whom computers, telecommunications and defence equipment are the primary sources of scrap.[2]

Hundreds of thousands of computer systems have been installed worldwide over the past decade. The pace of change in the computer industry is

---

[1] B. W. Dunning, Jr., Characterization of Scrap Electronic Equipment for Resource Recovery, *Proc. Sixth Mineral Waste Utilization Symp.*, pp.402–410, Chicago, May 2–3 1978.
[2] Subhash C. Malhotra, Trends and opportunities in electronic scrap reclamation, *Conservation & Recycling*, Vol.8, No.3/4, 1985.

so rapid that their lifetime is short before they are replaced by new and more efficient systems. Large, so-called 'mainframe' computers have a useful life of 8–20 years. Minicomputers, sold for domestic or recreational use, have much shorter lives but their distribution is more disaggregated.

7.4 CURRENT RECOVERY PRACTICE

Computer scrap, purchased from the user by a broker, or pre-processor, is manually stripped and segregated into categories such as back boards, printed circuit boards, and connectors. These then go to refiners, often in low labour cost areas, where the metals are recovered.

The operations at a typical reclamation facility handling all types of electronic scrap are very complex. Manufacturing scrap is subdivided into 22 basic product classifications. Old scrap falls into an even greater number of categories.[3]

A new method of copper recovery from printed-circuit boards involves leaching with aqueous $CuCl_2$ and subsequent reaction with aluminium to recover copper and $AlCl_3$.[4]

The United States Bureau of Mines has published details of a pilot plant for recovering precious-metal-bearing concentrates from shredded electronic scrap. It involves air-classification, wire-picking, magnetic separation, screening, eddy current separation and high-tension separation.[5] It was, however, recognised that hand-sorting was potentially beneficial in segregating easily recognisable components, such as an aluminium chassis and large components plated with precious metals.

Some 93% of the gold and 78% of the silver may be made to report to fractions representing only 26% of the total sample weight. Operating costs were estimated as from $241/tonne for a 5.4 tonne/day plant to $48/tonne for a 54 tonne/day plant. This represents processing costs only, and does not include a cost for the feed material or incorporate credits for the product materials.

7.5 CHANGES IN MATERIALS USE IN ELECTRONIC GOODS

By the nature of electronics scrap, its valuable components are used in small quantities, and are mingled with much larger quantities of valueless insulation. The pace of technological change and the rapidity with which electronic goods become obsolete are such that anticipated lifetimes are shortening. Thus it is reasonable to expect that since contacts will not need to have such a long life, manufacturers will be able to reduce the thickness of gold in them. The replacement of copper wire by printed circuits; will further reduce the value of electronic scrap in the future.

---

[3] Bernard O'Reilly, 'Communications scrap', Chapter VI.9 in C.L. Mantell (Ed.), *Solid Wastes,* New York, John Wiley and Sons, 1975.
[4] Minoru Hiromasa, *Jpn. Kokai Tokkyo Koho* JP 62 13,544 [87 13,544] (Cl. C22B15/08) Jan. 22 1987.
[5] Fred Ambrose and B. W. Dunning, Jr., Precious Metals Recovery from Electronic Scrap, *Proc. Seventh Mineral Waste Utilization Symp.*, pp.184–197, Chicago, Oct. 20–21, 1980.

This is additional to changes already being made specifically to achieve economies in materials use.

The pattern of materials change is completed by an increasing range of materials. It is claimed that at the electronic scrap processing facility of Western Electric, the incoming scrap contains 76 of the naturally occurring elements. Copper wire may be covered with aluminium, lead, or non-metallic materials. Steel wire may be coated with copper. Mercury relay switches; are a potential health hazard, as well as a potential contaminant.

Many of the materials now coming into use in the electronics and semiconductor industries are rare metals of very limited availability.[6] Production of metals such as gallium, germanium; and indium are currently less than 100 tonnes each per annum. The potential demand greatly exceeds the supply. Jacobson and Evans have analysed the supply situation for several of the materials which are of importance to the electronics industry.[7,8,9] The recovery of these metals is seen to be most desirable.

Scrap selenium contaminated with elements such as tellurium, arsenic and chlorine may be treated to recover 99·999% pure selenium by converting the alloy into a mixture of oxides.[10] Another patent treats granulated scrap alloy containing arsenic and selenium with aqueous NaOH; oxidation recovers the valuable constituents separately.[11]

Gallium may be recovered from residues containing both gallium and arsenic by treatment with chlorine gas to form crude gallium and arsenic chlorides. The purified electrodeposit is of 99·9999% purity.[12]

The reclamation of samarium from cobalt-samarium magnets is recognised as important in a context where samarium supplies are limited and likely to constrain future use of this class of magnet. Reclamation of samarium is as a double sulphate of Co and Sm, offering 95–100% recovery of Sm_2O_3 at 98·5% purity.[13]

7.6   THE BARRIERS TO RECYCLING OF ELECTRONIC GOODS

The value of materials in electronic goods is small compared with the added value. Unit values of recoverable metals are therefore also small.

[6] Chairman's Opening Address, *Metals for the Electronics Industry*, Commodity Meeting of the Institution of Mining and Metallurgy, Burlington House, London, December 3 1987.

[7] D. M. Jacobson and D. S. Evans, *Critical Materials in The Electrical and Electronics Industry*, Institution of Metallurgists, 1984.

[8] D. M. Jacobson and D. S. Evans, The Supply Economics of Germanium, *The Metallurgist and Materials Technologist*, Vol.15, pp.132–135, 1983.

[9] D. M. Jacobson and D. S. Evans, The Supply Economics of Gallium, *The Metallurgist and Materials Technologist*, Vol.15, pp.183–185.

[10] Santokh S. Badesha, U.S.Patent US 4,530,718 (Cl.75–121; C01B19/02), Jul.23 1985

[11] Kiyotoki Uehara *et al.*, Recovery of arsenic and selenium from scrap alloy, *Jpn. Kokai Tokkyo Koho* JP 63 30,826 [87 30,826] (Cl. C22B7/00 Feb. 9 1987].

[12] Shigeki Kubo, Method of recovering gallium from scrap containing gallium, Eur. Pat. Appl. EP 219,213 (Cl. C22B58/00), Apr. 22 1987.

[13] Hideo Koshimura, Recovery of Samarium from Scrap of Samarium-Cobalt Alloy with the Double Salt of Samarium Sulfate, *Kenkyu Hokoku — Tokyo toritsu Kogyo Gijutsu Senta*, (16), pp.113–8, 1987.

The barriers to the more widespread recovery of materials used in electronic equipment can be identified as:

(a) the small amounts of recoverable materials,
(b) the diversity of such materials,
(c) difficulties of collection, and
(d) the large amounts of associated worthless material.

The variety of recoverable materials used in electronic equipment is large, and their concentration is small. Characterisation, in order to identify the financial feasibility of a possible recovery operation, is difficult, complicated and intrinsically unreliable. There is considerable scope for a rational approach to the identification of components which contain worthwhile concentrations of scarce elements, so that they can be recovered without the need to process the associated *detritus*.

8

RECOVERY OF MATERIALS FROM TURBINE ENGINE SCRAP

8.1. INTRODUCTION

The materials to be considered under this heading are:

(a) Superalloys;
(b) Titanium alloys;

The economies and national defence systems of the industrialised countries depend on a number of imported raw materials. The nature of aviation, where engine and airframe manufacturers are always searching for more efficient engines, either in terms of thrust, fuel economy and eventually payload, makes these applications exceptionally intolerant of substitutions offering even marginally poorer performance. This is likely to make the materials less tolerant of contamination than would be the case in other applications.

The scale of cost of aviation applications must be considered. Since performance and safety are paramount, and since development costs are high and production runs low, economies of scale are not usually possible. In general manufacturing industry raw materials may typically account for some 40% of the total manufacturing cost. The so-called 'downstream effect' means that even a doubling of materials cost causes a much smaller percentage cost increase to the consumer. In aerospace applications raw material costs form a much lower proportion of total cost. A quantity of alloy costing perhaps £5 may be used to manufacture a blade sold to the airline for £300; this is a measure of the rigour of the manufacturing, testing and inspection procedures applied. It should also be noted that this care extends to an identification system that permits the blade manufacturer to trace not only the details of a blade itself but also the analysis and source of the materials used to make the alloy employed in it. Under such conditions the cost of the raw material becomes almost irrelevant.

There is a considerable bibliography of the critical and strategic elements of the problem, which need not be discussed here.[1]

The more efficient use of recycled foundry scrap or revert alloy presents an alternative to substitution as a means of conserving strategic raw materials and represents more effective utilisation of resources. In superalloys, however, a major difficulty arises from the inferior performance of revert alloys, which can result in hot tearing or in unacceptable levels

[1] John B. Wachtman, Jr., The nature of the critical and strategic materials problem. *Proc. 57th Meeting of the Structures and Material Panel of AGARD, 'Materials Substitution and Recycling',* Vimeiro, Portugal, (Oct. 9th-14th, 1983)(Agard Conference Proceedings No.356)

of microporosity. Hence, recycled material has generally been used in the less-critical applications.

8.2   CONTAMINATION IN SUPERALLOYS

There is scope for a better understanding of the relationship between casting conditions and microstructure in conventionally cast alloys, and for more comprehensive specifications for impurity elements.

Instabilities and uncertainties in price and long-term availability of certain elements which are essential to obtain good high temperature performance of superalloys make it desirable to use larger amounts of recycled material. In the aviation industry performance transcends price to an extent possibly unmatched elsewhere. However, much scrap is currently recycled within the industry to be used in the less-critical applications. The need to recycle arises, at present, less from shortages of component materials than from the large amounts of high-quality scrap regularly produced by individual foundries which may achieve yields of only 40%.

The major element concentration of recycled material is adjusted during remelting to meet the original specification. It is generally agreed, however, that the foundry performance of revert material is inferior to that of virgin. This results from variations in the level of trace elements such as nitrogen, oxygen, silicon and zirconium, which can enter the alloy from contact with the mould, cores and furnace environment.

An increased awareness of the harmful effects of elements such as Ag, Bi and Pb on the high-temperature performance of superalloys led to the establishment, in the 1960s, of specifications agreed between customer and supplier. They fixed acceptable limits of individual trace elements. Such specifications have, as more and better data have been made available, become more stringent. A further development has been the concept of Lower Reporting Limit (LRL), with the aim of indicating the presence of significant amounts of impurities within the limits set by the relevant specification. The downward trend in lower reporting limits, largely a consequence of better analytical techniques, is exemplified by the fact that the limit for bismuth has been reduced by an order of magnitude since 1975; between 1981 and 1984 the LRL for calcium has been reduced from 100 to 5 ppm.

Many elements are regularly analysed at concentrations above those of the LRL but, at the amounts normally detected, none even approaches the maximum specified values. Elements such as Ca, Sn and Zn are commonly agreed to be of little concern in control of product quality.[2]

The measures described provide a basis for the control of deleterious trace elements in;, but with the important exception of military aircraft

[2] P.N. Quested, T.B. Gibbons and G.L.R. Durber, Trace elements in superalloys and the implications for recycling, *Proc. 57th Meeting of the Structures and Material Panel of AGARD, 'Materials Substitution and Recycling'*, Vimeiro, Portugal, (Oct. 9th-14th, 1983) (Agard Conference Proceedings No.CP-356, Paper No.19)

applications, national specifications for superalloys do not usually include comprehensive limits for impurities. For example, for an IN100 alloy, the most rigorous specification with respect to impurities generally recognised as harmful, viz. AFNOR NK15CAT, gives limits for Ag, Bi and Pb only. The corresponding BS HC204 specification includes limits for Ag and Pb only, and the latter is fixed at twice the AFNOR level. These specifications apply to both virgin and revert heats, and permissible Si and Zr can readily be met during scrap remelting (known in the industry as 'reverting'). No limits are laid down for control of gases such as nitrogen and oxygen. Nitrogen, in particular, can seriously impair the performance of Ni-based alloys.

The impurity content of virgin melts can in principle be controlled by suitable choice of raw material and processing route. Some constituents can themselves introduce contamination. Of the major raw materials used in superalloy production, chromium is the major source of contamination, especially of sulphur and phosphorus.

Analysis of about twenty casts of Mar M002 showed the impurities to be, for the most part, well below the specified limits. Similar conclusions were reached for the casts of virgin and revert IN713 LC. Some of the differences in trace content are attributable to variations in major element composition.

Revert alloys are generally characterised by higher nitrogen and, in some cases, silicon content. The higher Cr alloys, viz. IN939 and IN738, generally have higher nitrogen concentrations. Some control of nitrogen level is possible in revert casts by careful scrap selection and by removing the surface layer from recycled castings, since nitrogen contents of >100 ppm have been measured at the surface of cast superalloys.[3]

Silicon contamination especially in small parts and those cast by directional solidification, results from interaction with moulds and cores; modern analytical techniques reveal that revert Mar M002 can contain at least six times as much silicon (0·06%) as virgin (>0·01%).

8.3   EFFECT OF RECYCLING ON MICROSTRUCTURE AND PROPERTIES

Castings from recycled superalloys are characterised by higher levels of microporosity in revert alloys; compared with virgin heats, impairing stress-rupture properties. Consequently, the use of revert has tended to be restricted to castings with good feeding characteristics. Alternatively, hot isostatic pressing (HIP); may be used to close the pores. There is evidence that poor foundry performance is linked with the high gas contents typical of revert casts.

Carbon can also influence the amount of microporosity in superalloys, but there is no evidence of significant variations in the carbon content between virgin and revert melts.

[3] P.N. Quested *et al.*, *loc. cit.*, No.19.

8.3.1 Implications for foundry practice

Two principal alternatives exist to facilitate the use of recycled Ni–Cr base alloys; in complex castings for gas turbine applications:

(a) Modification of the alloy composition during remelting to restore the minor element content to that of the virgin alloy.
(b) Control of the investment casting conditions to enable revert material to produce castings whose soundness matches that of the virgin material.

In view of the difficulties of controlling oxygen and nitrogen content, the second alternative appears the more attractive. Indeed, certain foundries claim to produce sound castings in complex shapes with high revert ratios.

Detailed studies on the relationship between casting conditions, solidification structure and microporosity have been carried out for directional solidification;[4] however, these data are not available for conventional casting.

8.3.2 Cross Blending

The technique of 'cross-blending', which is commonly used in the secondary materials industries, may provide an opportunity for avoiding some of the foundry problems associated with the high nitrogen content of revert alloy. The case in point is the use of scrap IN713, which contains niobium but no cobalt, with scrap IN100, which contains cobalt but no niobium, to produce a melt of IN738, which contains both elements. The advantage from the point of view of nitrogen content is that, even in virgin melts, the permitted nitrogen content of the high-Cr alloy IN738 is substantially greater than that for revert IN100. In this way, valuable scrap could be recycled without prejudicing foundry performance. The development of this technique should be encouraged.

8.3.3 Recycling practice

The range of nickel alloys includes Ni–Cu (Monel), Ni–Fe (Invar), Ni–Fe–Cr (Incoloy) and Ni–Cr–Co–Mo (Nimonics, of which there are several). Typically, only 50% of the material entering a superalloy manufacturing plant leaves as prime product. The other 50% arises as in-house scrap.

The critical nature of the applications for some of these alloys is such that it would be disastrous were they to become mixed. For example, the Nimonics are particularly sensitive to copper. The aim is to avoid contamination during melting. This involves the scheduling of melting such that the furnace lining is used for alloys *in a given order;* a given lining cannot be used for Nimonics after having been used for Monels. Subsequently, all alloys have to be kept completely separate. The deliberate

---

[4] J.R. Stephens, NASA Technical Memorandum 82852, 1982

mixing of scrap is a dismissible offence at some, perhaps most, alloy manufacturers. Each batch of alloy is accompanied at all stages by a card; only one type of company scrap label is permitted. To prevent material falling out of the scrap hoppers, with the risk that they might be returned to the wrong container, hoppers are customarily loaded to only three-quarters capacity.

Some alloys are produced only on an occasional basis, and their composition is such that the scrap may not be assimilable into other grades. This means that the production of such grades must be done in small batches so that the scrap can be reassimilated into the next melt. Segregated scrap arises in various forms, ranging from massive pieces to finely-divided swarf, and the metal warehouse is organised with the aim of keeping these separate. Each large piece is individually labelled to verify the card, and may be cleaned by degreasing or shot-blasting, as appropriate. Shot-blasting is necessary to remove the glass lubricant from the extrusion process. The variable composition of swarf is such that the computer may need to average the composition of turnings over the last three months. Parcels of scrap are kept to less than 2 tonnes because the aim is to use an entire parcel in one melt.

Because the composition of turnings is always uncertain, each alloy will have its individual limitations on the amount that may be used. Some alloys permit no turnings to be included.

8.3.3.1 *Bought-in scrap*
Consumer companies may be unused to the discipline of scrap segregation. Turnings are especially likely to be contaminated. Within a superalloy manufacturing plant, a machine will be cleaned up after one batch and before the next, but this is unrealistic for external scrap. Alloys of low melting-point may arise as contaminants because of the tendency to use low melting-point material, such as Wood's metal, for holding pieces of Nimonic during machining operations. The melting analysis and the holding of materials of unusual composition for eventual use is not viable, particularly at times of high interest rates. Bought-in scrap may sometimes come in as part of a transaction made for other, more compelling, reasons; it must be pre-melted and analysed, with an inevitable increase in cost.

The hierarchical approach is adopted. Alloys which cannot use their own scrap are unpopular. Since silicon pick-up can result from ladles the refractory may need to be changed or the product may be vacuum-melted and cast. This eliminates silicon pick-up because there is then no transfer ladle. Vacuum-melted and cast alloys use only vacuum-melted and cast scrap. Nonetheless, there are still some possibilities for scrap transfer.

Low melting-point materials need contribute only a few parts per million (ppm) of certain impurities, such as lead and zinc, for damage to result to the properties. They can be removed by vacuum-melting, but the process is time-consuming. Tin cannot be removed in this way.

Superalloys are always vacuum-treated at some stage, but the standard melting time may have to be increased if impurity levels are high.

Undoubtedly, given its frequency of occurrence, lead is the most troublesome impurity. Bismuth, however, although less frequently encountered, is actually a more potent contaminant. Since it cannot be tolerated to more than 0·5 ppm it is important to eliminate swarf produced by machining in Wood's metal.

8.4 Titanium

Titanium alloys may broadly be divided into:

(a) Commercially pure titanium
(b) Specialised alloys

Commercially pure titanium is used principally in the chemical industries. Some is used in air frames and engines. The commonest grade is Ti–6Al–4V. Volume production is concentrated in commercially pure titanium, which is most conveniently processed on plant closely related to steel-rolling plant, and most heavily developed in Japan. Normally Ti–6Al–4V is a single alloy, but companies may make several grades, each carrying its own designation within that grade.

In the aerospace sector, applications are subdivided into two main categories: 'rotor' grade, (the more critical rotating components), and 'non-rotor' grade (parts which do not rotate and are relatively lowly stressed). Within each main category there is a spectrum of grades. Titanium is invariably double-melted, and sometimes triple-melted. Certain rotor grades require control not only of the properties and the chemistry but also of the method of manufacture with, for example, control of the forging schedule. Products, and the method of manufacture, are normally approved by the end-user, involving costly testing. Hence, a decision to change something nominally as simple as the ingot size would involve making components from the new ingot size and subjecting them to spin and all other appropriate tests. Such a testing schedule is extremely costly, and could involve millions of pounds sterling. Therefore, changes cannot be undertaken lightly.

In the late 1970s, titanium enjoyed a great increase in popularity. In the West, the output was limited by the availability of new material, especially in the U.S.A. This was the point at which, of necessity, large-scale recycling operations started. New processing methods were developed for scrap, whose use is viewed more sympathetically in the U.S.A. than in the U.K.

The large titanium companies are interested in scrap recycling for the usual financial reasons but, for reasons connected with quality, there is much suspicion involving scrap. Only now are the alloys developed in 1979/80 beginning to enter service in the air. Typically, a large titanium company will divide scrap into internal and external grades. Internal scrap comes from areas where direct control can be exerted over its quality, e.g. where it can be certified to be free of heavy metal contamination. It is axiomatic that every titanium melter is now well-practised at processing such material. The real doubt arises with external scrap.

Two principal types of problem arise:

(a) Low-density inclusions (LDI) — interstitially stabilised defects.
(b) High-density inclusions (HDI).

8.4.1 Swarf

8.4.1.1 *Internal scrap*

Scrap can broadly be divided into metallurgical scrap and processed scrap. Internal operations permit quality control on scrap arisings, e.g. in factories where it is possible to ensure that no heavy metals are being processed. Cutting tools containing tungsten carbide contamination are a risk, so such plants may stipulate that only magnetic tools be used so that cutting debris can be removed and recovered on a 100% inspection basis. The amount of recycled swarf included in ingots is governed by the oxygen content. The machined skin of first-melt ingots is not recycled but sold. Metallurgical scrap is material that has for some reason failed, whereas processed scrap, including offcuts, is material that has been found to be acceptable.

8.4.1.2 *External scrap*

The processing of swarf is limited to crushing and washing. It must then be tested.

The first problem is identification on a cost-effective basis; hence, swarf is preferred from a source that does not process potential contaminants, or where adequate cleaning is known to be carried out. On a commercial basis, separation of the individual titanium alloys in swarf is impossible; however, it is possible on large pieces.

Smelters in the U.S.A. now use swarf processed by, for example, the fluidised-bed technique, but this will guarantee only minimum standards. Final inspection must be by X-ray.

There is a potential problem of scale of operation that transcends ordinary economies of scale. Large manufacturers of aviation products may be able to justify dedicated machine shops for individual titanium alloys; this may be impossible for smaller manufacturers.

8.4.2 Massive scrap

8.4.2.1 *Internal massive scrap*

In-house scrap raises concern about the presence of fragments of machine-tool embedded in material which ultimately will form scrap. For high-value applications, such scrap must be pickled and inspected.

95

Generally, flame-cut surfaces are not liked because of the possibility of the pick-up of interstitials; such surfaces may need to be shot-blasted and pickled, or simply cut off mechanically.

8.4.2.2 *External massive scrap*

Control of impurities is via elimination, not reduction. There is considerable concern over scrap components, i.e. any object that has been machined, in view of the likelihood that machine-tool dbris may have become embedded in the surface. Welds involving a tungsten electrode are unpopular. The only really acceptable method of welding is electron beam. Moreover, components are sometimes built up, which can introduce impurities. Even components such as so-called 'clappers' or 'snubbers' are faced with tungsten. Keyways are sometimes faced in areas which are difficult to inspect. It is not worthwhile processing old blades because so much of the added value of blade material resides in the implicit assurance of quality associated only with virgin material.

This is true for all manufacturers.

Of the remaining material, forging scrap and off-cuts are small in volume. One problem for the reclaimer is the need to assume that the original material was in specification. It is not feasible to X-ray everything. Another distinct problem is that of new, i.e. unknown alloys. Each major titanium manufacturer has lists of the alloys of its competitors, whose specification is well-publicised through a desire to make the material well-known and therefore popular. Other countries, especially the Eastern-block countries, are less obliging.

One approach to the scrap-processing problem is to design systems, e.g. ferrofluids, specifically for separation of high-value materials, or to reduce the costs of X-ray inspection. Another is to develop refining procedures using non-consumable electrodes, e.g. half-melting, which relies on the sinking of the HDI into the stub. Electron-beam melting evaporates some contaminants, e.g. aluminium.

All titanium is melted twice. Additional melting to make compositional corrections incurs insupportable losses considered in financial terms. Objectionable heavy metals include hafnium, molybdenum, niobium, tungsten and zirconium. Some of these, e.g. zirconium, are easy to detect, whereas a particle of tungsten carbide, 0·4 mm (0·015″) in diameter is not. A novel process, the so-called 'package melting' method facilitates remelting of swarf, chips and massive scrap together. It is said to hold cost advantages over other melting techniques.[5]

The average product yield is 65%. Of the 35% lost, some 18–20% ends up as massive scrap, 10% as swarf, and the rest is lost. There are significant losses in scrap processing. Some 25–30% of input is process scrap, but yields are very sensitive to product mix. Sheet is extensively pickled; forging scrap, by contrast, is produced as massive material,

---

[5] Tsutomu Oka and Yoshiharu Mae, *Tetsu to Hagane,* **73** (3), pp.520–7, 1987.

which requires a different approach. The figures given above assume no external scrap. Losses are also sensitive to product mix by alloy.

Competition is increasing in the titanium field. Some manufacturers rely heavily on one customer, which may have its own inspector on the manufacturer's premises, to ensure that all quality control standards are met. Worldwide, titanium is in oversupply. It sells for around £6/kg.

In the industry, a major problem is seen in the incidence of second-generation alloys, i.e. alloys made using scrap that itself contains scrap. In the U.S.A. the market is large enough to support producers of both high- and low-quality alloys. Some trace elements which are very uncommon, e.g. yttrium, are nonetheless sufficiently troublesome to make it worthwhile to analyse for them.

A new, and potentially dominant, factor is the concept of product liability. Ultimately, the risk seems likely to lie with the smelter.

8.5   THE BARRIERS TO TITANIUM RECYCLING

The main problems that exist for titanium scrap recycling in the U.K. are:

(a) Quantities are too small, relative to aluminium, to facilitate extensive operations.
(b) Improvements are needed in melting and refining.
(c) More efficient cutting tools are needed. In particular, borides are undesirable.

The scrap market is particularly volatile, in view of its small size. It can be affected by comparatively small influxes of material.

8.6   CHANGES IN MATERIALS USE IN AVIATION

Many new problems have arisen with the introduction of coatings. Generally, all metallic and many ceramic coatings preclude the eventual use, as scrap, of the material to which they are applied. Many ceramic coatings are of a proprietary nature, whose manufacturers will not disclose their composition.

Composite materials, designed to obviate the modulus problem in titanium, will present other problems. Powder-formed alloys introduce chloride contamination, presenting difficulties in the welding of fabricated sheet.

8.7   DESIGN CHANGES TO FACILITATE RECYCLING

From the foregoing, it is clear that compared with other industries, the manufacturers of titanium-based and nickel-based alloys used in gas-turbine engines go to extraordinary lengths to prevent contamination of their alloys. The scope for reassimilation of external scrap — which includes post-consumer scrap — in alloys destined for aviation use is very limited indeed.

It is difficult to envisage any circumstances that could induce the designers and manufacturers of articles of such high added value, or with such stringent safety-related specifications, to design with ultimate recovery in mind.

There are, however, uses for titanium other than in the metallic form or in aviation. There is scope for the development of lower-specification titanium alloys, where impurities would be less deleterious than in aerospace.

In 1983, only about 1·5% of titanium consumption in the U.S.A. was as metal, and less than half of that was in aerospace. Over 98% went into dissipative applications, principally potential use of scrap in paint (47%) and paper products (26%).[6] Titanium oxide pigments for paint must be low in iron. Other consuming industries impose their own specification. It seems, however, highly likely that most or all of the metal recovered from the aerospace industry could, in principle, be absorbed in pigments or chemicals if the titanium industry itself cannot reassimilate it.

[6] Langtry E. Lynd, Titanium, *Mineral Facts and Problems,* 1985 Edition, Washington, D.C. U.S.Department of the Interior, Bureau of Mines, 1985.

9

PRECIOUS METALS

9.1   RECOVERABLE PRECIOUS METALS

The uses of precious metals are becoming more diverse and, largely as a result, it is also becoming more difficult to reclaim them. Some aspects of their use have been discussed in the chapters dealing with road vehicles (Chapter Five) and electronic goods (Chapter Seven).

Some 40% of silver consumption in the U.S.A. goes into photographic products. The metal is recoverable from them and, indeed, large laboratories routinely treat their waste fixing solutions. To collect from the ultimate consumer, i.e. the general public, however, presents formidable problems.

The platinum group of metals (PGM) comprises platinum, palladium, rhodium, ruthenium, iridium and osmium, which are among the scarcest of the metallic elements.[1] This group has become critical to industry because of chemical and physical properties that are unparalleled. In many of the industrial applications which account for some 95% of yearly consumption of platinum and palladium in the U.S.A., and some 90% of PGM consumption worldwide, there are no realistic substitutes for these two metals.

The metals are not only scarce but also very locally distributed. Some 81% of the world's known platinum-group reserve base is located in South Africa, and another 17% is in the USSR. South Africa accounts for 85% of the so-called Western World production of PGM. This very restricted distribution makes world supplies unusually dependent on conditions within one country; in January 1986, when a strike began at Impala Mines, South Africa's second largest producer of PGM, there was an immediate sharp rise in the price of platinum.

The PGM metals are used, among others, in the industries concerned with chemicals, electrical equipment, glass, jewellery, and petroleum refining.

Among the other precious metals, the trend towards miniaturisation is reducing unit metal use, and substitution of palladium for gold and silver in contacts and connectors has significantly reduced the value of scrap components. Aspects of this problem have already been discussed in Chapter Seven, 'Electronics scrap'.

[1] A.D. Owen, The Demand for Platinum, *Resources Policy*, Vol.13, No.3, pp.175–188, September 1987.

Ironically, the single factor that makes the precious metals so useful, i.e. their nobility and resistance to oxidation and corrosion, confers on them their ability to be useful in the very small concentrations in which they are used in contacts, and in which form they are so widely dispersed. The artefacts employing them, e.g. hand-held calculators, are resources of low grade, in which small quantities of valuables are associated with much larger quantities of valueless detritus. Further, the vast range of types and classification of materials requires special sampling and processing.[2]

One very large use of platinum and palladium does, however, offer the prospect of easy collection. This is the catalytic converter in cars, mentioned briefly in Section 5.12. Before 1974, platinum-group metals were not used in the automotive industries except for points and sparking plugs. In 1974, however, automotive applications accounted for 37% of U.S. demand for platinum and 17% of palladium. By 1979, the figures had risen to 57% and 20%, respectively;[3] by 1983, automotive platinum demand had risen again, to 64% but automotive rhodium use had eased somewhat, to 18·7% of total consumption.

The quantities used per so-called three way catalyst (a type which simultaneously deals with hydrocarbons, carbon monoxide and oxides of nitrogen, NO_x) are small: 0·05 oz. Pt, 0·02 oz. Pd, and 0·005 oz. Rh are typical figures. However, the numbers of such converters make them major consumers of PGM.[4] The sheer quantity of PGM consumed is not, however, the only resource problem associated with the converter. The ratio of platinum: rhodium employed is not the one which occurs naturally in minerals (see Section 5.12). Depending on engine design, the average ratio of rhodium: platinum used in the converter is 1:5.74. Since the average mine ratio, based on the Merensky reef, is only about 1:19, there are fears of a potential future shortage of rhodium, with price rises as a result. Redesign of the automotive catalyst is not currently possible because the Pt:Rh ratio in it is, with current knowledge, the most favourable for catalysis. Research and development in catalyst design is aiming at reducing the proportion of rhodium and increasing that of the much cheaper platinum.

In 1979, platinum production worldwide was more than 14 × rhodium production. By 1983, this ratio had declined hardly at all, to 13·5. In some deposits, the Pt:Rh ratio is 19:1. Therefore, an automotive ration of 5·74:1 is distorting the marketplace for platinum-group metals. However, the development of the UG2 Reef, with a Pt:Rh ratio of 5–6:1, should partially alleviate the problem, at the cost of placing greater dependence than ever on South Africa as a source of PGM.[5]

[2] Precious scrap increasingly complex, *Metal Bulletin*, May 21, 1985.
[3] James H. Holly, Platinum-group Metals, *Mineral Facts and Problems*, Washington, D.C., U.S.Department of the Interior, Bureau of Mines, pp.603–706, 1980.
[4] J. Roger Loebenstein, Platinum-group Metals, *Mineral Facts and Problems*, Washington, D.C., U.S.Department of the Interior, Bureau of Mines, pp.595–616, 1985.
[5] *Platinum 1987*

The economics of recovery from catalytic converters; of PGM from catalytic converters in 1987 suggest that the break-even point, at 1987 costs, to make recycling viable is around \$350–400/troy oz. for platinum. When, in 1980, the price rose to more than \$1,000/oz., each converter was worth approximately \$16. At \$240/oz., as in 1985, the converter lost interest as a source of PGM.[6] Recovery is not straightforward, since the metals are dispersed in minute doses on ceramic pellets or on a ceramic honeycomb designed to achieve the maximum possible exposure of catalyst surface area to the exhaust gases. A converter weighing some 12 kg might contain less than 3 kg of ceramic, of which only 1·4 g is precious metal.[7] However, at suitable price levels for platinum, this may represent a worthwhile return from PGM recovery in recoverable values, in addition to possible revenues from the sale of the recovered steel casing and the ceramic carrier.[8]

As in electronic goods, the value of the precious metals is so small compared with the added value of their application that it is difficult to envisage redesign being carried out for the purposes of recyclability. The principal route to improved recovery is seen as improved collection of the appliances and electrical goods in which precious metals are used.

New processes for precious metals recovery apply leaching, electrolysis, and selective dissolution of the non-ferrous metal substrate to scrapped engineering and electronic devices, precious-metal-containing catalysts, and non-ferrous metals with precious metal plating.

Research to identify possible substitute materials for PGM in automotive catalysts has shown that certain Group IVB elements (Hf, Ti and Zr) show some promise.

[6] Anon, Auto Catalytic Converters: A "Mini-mine" on wheels, *Phoenix Quarterly*, Vol.19, No.2, pp.8–10, Summer 1987.
[7] Salvaging Auto Catalysts', *Business Week,* Dec.18 1978.
[8] Sebastian P. Musco, Reclaiming of Precious Metals from Automotive Catalytic Converters, *Proc. 3rd International Precious Metals Conf. of the International Precious Metal Institute*, Chicago, May 9 1979.

10
HAZARDOUS WASTES

10.1 INTRODUCTION

Along with those of many other industries, the materials recycling companies are now obliged to comply with disposal regulations that become ever more stringent. The enactment and more rigorous enforcement of environmental regulations in the environmental legislation; U.S.A. have already resulted in reduced activity in reclamation and, necessarily, in increased disposal of certain residues which contain materials that it might be thought desirable to reclaim. Since proper, environmentally-sound disposal is an expensive procedure, a decision not to recycle is likely to lead to uncontrolled disposal.

The environmental movement has had a marked effect on recycling patterns, with emphasis now being placed on what are termed 'Hazardous materials'. Uncertainty over materials or end products which might be presented for processing but which might, later, be categorised as hazardous is already leading some scrap processors to refuse to handle certain categories of scrap.

Recent American legislation on absolute product liability has sharply focused attention on the problems faced by reclaimers who have to dispose of residues containing a number of hazardous materials now widely recognised as particularly toxic.[1] A comprehensive survey of these is beyond the scope of this study. However, they include:

Arsenic:
Barium:
Cadmium: This is used as a protective coating on selected bolts
               used in automotive engineering, and as a colouring
               medium in household appliances. Its use is banned in
               certain countries, notably Sweden and Denmark.[2]
Chromium:
Lead: Lead is still used in motor fuel, despite moves to
               phase it out in favour of lead-free grades.
Mercury: Used in thermal switches

---

[1] See, for example, See No Evil: Can $100 Billion Have 'No Material Effect' On Balance Sheets?, *The Wall Street Journal*, May 17 1988, p.1.

[2] Anon, Design for recyclability will help solve problems of hazardous waste, *Materials Reclamation Weekly*, Oct. 12 1985

Selenium:
Silver:
Sodium azide: This substance, the active constituent in passive seat belts or air bags, introduces the risk of uncontrolled explosion in shredders, and possible mutagenic and carcinogenic dangers.

Coated metals might be thought to pose an environmental threat if the coating is of a so-called heavy metal and if the concentration of such metals in processing residues reaches a critical level. Hence, the residue can be an hazardous waste.

Hitherto, the approach to the hazardous waste problem has been to try to manage the wastes only after they have been seen to be a hazard. Although a necessary response, it is only a short-term solution and an extremely expensive one, designed to limit environmental damage. Ideally, the need is to address the question of the continued introduction of such potentially hazardous materials.

A significant, if not the principal problem, concerns the actual *modus operandi* of these elements. Clearly, one may identify certain chemical species as toxic and may determine the lethal dose in a healthy adult of standard weight. What is not known is the long-term effect under less than standard circumstances. It is conceivable that long-term exposure — a week, month or year — by a child or less-healthy adult might cause no immediate or conspicuous symptoms but yet result in severe or crippling illness, or death, decades later.

The problem does not end with the materials of construction. The potential for the contamination of processing facilities has obliged North American scrap processors to reconsider the advisability of handling barrels, drums and other containers as recyclable items. Some scrap purchasers have placed a total ban on such containers which might have contained hazardous wastes, such as pesticides. Others accept only containers from known sources, where they can be confident that appropriate cleansing procedures have been followed.[3]

It is an irony that some industries will refuse to accept scrap produced from certain of their own products. For the reasons given in Chapter Four, galvanised steel may be unacceptable as scrap at the very steelworks that produced it. High-copper steel scrap, produced from corrosion-resistant pipe, is likely to meet similar resistance.

It is not, therefore, surprising to discover that the scrap industries are beginning to refuse to purchase or handle certain types of secondary material, because of the potential for the generation of hazardous wastes and because of the possible legal implications. An arrangement whereby all materials producers are obliged to liaise with their customers and with the reclamation industry could only be beneficial in the question of hazardous wastes.

[3] Anon, ISIS Cautions Pesticides Wastes Disposers, *Phoenix Quarterly*, **18**, No. 1, Spring 1986.

The potential dangers of these elements which, with the exception of lead, are used in small amounts, are such that none of them should henceforth be used in any application whatsoever without first considering:

(a) hazards of its use in service,
(b) problems in recovering the hazardous materials at the end of the useful life of the product, and
(c) potential for hazardous residue generation.

The Institute of Scrap Iron and Steel, Inc., The (ISIS),[4] as one of the major representatives of the scrap industry, has set out a proposed action plan involving:[5]

(a) Development of an inventory of products containing U.S. Environmental Protection Agency (EPA) EP toxic heavy metals. Under this phase of the programme, the EP list of eight heavy metals (Ag, As, Ba, Cd, Cr, Pb, Hg, Se) would be the basis for establishing which products have the potential for contaminating recycling residues.

This list is not, of course, exhaustive, since it omits extremely toxic materials such as beryllium and its oxide.

(b) Consideration of how best to handle potentially hazardous materials already in service.

(c) Definition of the best means of encouraging the future substitution of non-hazardous materials for the hazardous ones used now and in the past.

(d) Consideration of policies for dealing with problems posed by materials for which substitution seems unrealistic.

The 'Grey List' and the 'Black List' of the EEC sets out similar warnings for Europe.

The key factor in the possible elimination of potentially hazardous materials by substitution is an economic one. Possible measures to influence the economic decision include tax credits, accelerated amortisation, reductions in rates (property taxes) and preferment in Government contracts. Application of some or all of these measures would be seen to reverse the financial justification for continuing to use an apparently cheaper, but hazardous, material rather than a more expensive, safer alternative. Much of the legislation to achieve these ends is already in place.

---

[4] Now amalgamated with NARI (National Association of Recycling Industries) to form the Institute of Scrap Recycling Industries, Inc.
[5] Anon, Design for Recycling, *Phoenix Quarterly*, **18**, No. 1, Spring 1986.

In the U.S.A., the law known as 'Superfund' provides for control under the headings of:

(a) strict liability,
(b) joint and several liability, and
(c) no statute of limitations.

A comprehensive examination of the legal aspects of 'Superfund' is not appropriate here. Briefly, 'strict liability' signifies that ignorance is no defence and neither is the absence of causality. Thus, liability can attach to alleged contamination which occurred before, for example, there was any suggestion that polychlorinated biphenyls (PCBs); might be injurious. 'Joint and several liability' signifies that a single traceable shipper or supplier of a single contaminating item among millions of other barrels from non-traceable sources could be held solely responsible for an immensely costly cleanup operation. The absence of a statute of limitations means that a company which salvaged and processed a PCB-containing transformer in 1940 is still liable for any injury that traces of the PCBs might be alleged to cause in 1988.[6] The literature is now beginning to feature reports of investigations under Superfund; the alleged contaminations usually took place many years or decades ago.

In the U.S.A., the Institute of Scrap Iron and Steel, Inc. has published guidelines regarding materials whose processing should not be undertaken without due consideration. The eleven items in question must be removed from auto hulks at the processor's gate, or the entire load will be rejected. This list includes batteries, petrol (gas) tanks, extra silencers (mufflers) and tailpipes, closed containers, and dirt and débris; it comprises an attempt to ensure that the shredder *detritus* remains non-hazardous.

The above list relates specifically to shredders. A further 25 materials are listed by ISIS as requiring informed consideration by all processors prior to acceptance. Wire and cable are thought to create a lead threat, white goods a cadmium threat, and hospital scrap — as in the case of the caesium-containing cylinder abandoned in Sao Paolo — a threat of radioactivity. Drums are not handled if there is any possibility that they might have contained hazardous substances, and new methods and technology are being developed to deal with the disposal problem as more scrap processors refuse to purchase drums of unknown origin.[7] Transformers filled with polychlorinated biphenyls (PCB), which have been held — possibly on slender evidence — to be carcinogens, are treated only if they are guaranteed to be empty.

The implications for the scrap recycling industries are serious and profound. As an example, one may cite the disposal of lead-acid batteries. In the U.S.A., lead at concentrations of >5ppm constitutes a hazardous waste; hence, a decision to process lead-acid batteries or not is a risk calculation.

[6] Herschel Cutler, Mandate for change, *Phoenix Quarterly*, 18, No.4, Winter 1986–87.
[7] Anon, Drums are Marching to a New Drummer, *Phoenix Quarterly*, **18**, No.2, Summer 1986.

The standards which apply to spent batteries are ambient air quality standards for lead, Government (OHSA) standards for lead in the place of work, water quality standards for smelters and battery plants, and regulations for the storage and handling of hazardous waste.

In 1978, the ambient air quality standards for lead were set at 1·5 microgrammes/m^3, averaged on a quarterly basis for battery manufacturers and for primary and secondary lead smelters. The capital investment in plant improvement to meet this standard was estimated to be equivalent to 8·8c/kg (4c/lb), and had to be met by 1988. More recently, a more stringent standard of 0·5 microgrammes/cubic metre has been proposed.

Regulations of 1984 and 1985 classify spent lead-acid batteries as hazardous wastes. They provide *inter alia* for regulation of any business generating more than 100 kg of batteries per month. Storage of batteries on-site qualifies a processor as a land-disposal site, with additional registration formalities, and imposes a need for permits to continue operation.[8] The corrosive acid presents a second disposal problem.

The ramifications are considerable. The battery was, in the past, a valuable item that seldom, if ever, ended up in a landfill. Following enactment of the legislation, battery recycling has fallen from 90% of all discarded batteries *c.* 1980 to around 60% in 1986. Of some 650 000 tonnes of battery lead available for recycling in 1985, only 377 000 tonnes was actually reclaimed.[9] Processors are refusing to accept batteries. Car dismantlers are resorting to concealment to dispose of them; in one known case, 47 used batteries were discovered in a single car hulk. Moreover, landfills are not designed to contain the sort of pollution introduced by discarded batteries.

Further back in the lead production chain, the period 1983–1985 has seen a fall from 80 to 23 secondary lead smelters in the U.S.A. Most of the 57 closures have resulted from financial inability to meet environmental regulations. They accompany a fall in secondary lead-smelting capacity from approximately 1·5 million tonnes in 1981 to about 900 000 tonnes in 1984. Of some 300 operating battery breakers in the late 1970s, only five remained active in the entire U.S. by 1986.

The use of other types of battery presents further health hazards from the known toxicity of the mercury which is used in many patterns. Spent batteries of the manganese dry-cell type may be treated to recover manganese, mercury and zinc.[10, 11] The small size of such batteries makes it likely that they will be discarded through the refuse stream and will be

[8] Putnam, Hayes & Bartlett, Inc., The Impacts of Lead Industry Economics on Lead Battery Recycling, Prepared for the Office of Policy Analysis, Environmental Protection Agency, June 13 1986.
[9] Anon, Dead Batteries — A Negative Charge to the Environment, *Phoenix Quarterly*, **18**, No.2, Summer 1986.
[10] Hiroshi Iemura and Masaki Fujiwara, Zinc Recovery from Spent Batteries, *Jpn. Kokai Tokkyo Koho* JP 61,234,981 Æ86,234,981Å (Cl. B09B5/00, Oct. 20 1986).
[11] Byung Ha Yoon and Dai Ryong Kim, Recovery of Manganese from Electrolytic Zinc Anodic Slime and from Scrap Dry Cells, *Kumsok Pyomyon Choli*, 19 (1), pp.13–19, 1986.

either incinerated or enter a landfill, where corrosion will later allow them to leak heavy metals.

The advantages of such batteries are many, and it seems unlikely that they will be abandoned on environmental grounds. However, there would seem to be good reason for attempting to rationalise their distribution and collection procedures.

Other types of arising present not only a contamination risk in the eyes of the law, but a physical risk to the processor. Such a case is the unspent air-bag canister.

An inescapable result of environmental legislation of this type is that recyclability must be incorporated at the design stage. Very toxic materials and very dangerous components must be substituted by non-hazardous alternatives. If they cannot be replaced, they should be designed for easy identification and removal.

11
DESIGN FOR REMANUFACTURING

11.1 INTRODUCTION

When a so-called 'durable' product, such as a machine, vehicle or appliance, reaches the end of its useful life, it is usually disposed of in a landfill or is reclaimed for the sake of its raw materials. With either course, the costs associated with collection and operation of the landfill, or the costs of shredding, sorting, and melting reclaimable materials may well exceed the direct financial benefits of the operations. Thus, they may well not be carried out, and the durable may simply be discarded, possibly by illegal tipping.

In any single case, there is often some removal of still-serviceable spare parts, and this underlies the process of emanufacturing. This may be defined as the restoration of used products to a condition which, if not precisely as-new, has performance characteristics which approximate to new. It applies a series of industrial processes to worn-out or discarded products to disassemble, clean, and refurbish usable components, provide new parts where necessary, and reassemble and test. For many years, such rebuilding of unserviceable automotive electrical equipment has provided the consumer with a practical and economical alternative to the purchase of items manufactured from new components.[1] Typically, a factory-rebuilt starter motor for a popular type of family saloon might offer the purchaser a saving of 30% on the price of a brand-new unit.[2] A study on the remanufacture of chain-saws in the U.S.A. concluded that the operation could be profitable if the volume could reach 2000–2500 units per year. A volume of 25 000 per year would be highly profitable; such a volume would still be less than 2% of the average annual sales of chain-saws in the U.S.A.[3]

There is a substantial difference, usually about two orders of magnitude, between the retail cost of a finished product and its eventual value as scrap materials. It has long been recognised that benefit accrues to the consumer through an extension of product life or, alternatively, through the identification of a secondary use.

[1] Robert T. Lund, *Remanufacturing: The Experience of the United States and Implications for Developing Countries,* World Bank Technical Paper No.31, Washington, D.C., The World Bank, 1984.
[2] T.R. Cox and M.E. Henstock, An energy analysis of the reconstruction of a car starter motor, *Proc. Third Int. Conf. on Energy Use Management,* Berlin, Oct. 26–30 1981.
[3] Lynn Bollinger, Joel Clark, Richard L. Frenkel, Ronald Grand, Richard M. Kutta, Robert T. Lund and Floyd R. Tuler, *Energy Recapture through Remanufacturing. Final Report of Pre-Demonstration Study,* Cambridge MA, Massachusetts Institute of Technology, The Center for Policy Alternatives No.81–13, Jan. 1981.

Electrical units used in vehicles are assemblies of parts, most of which may be replaced when worn out or broken. The service life of such a unit can, in principle, be extended indefinitely. In some cases, rebuilding to extend product life may be necessary because the equipment is no longer in production, or because currency restrictions preclude the import of replacement units or spare parts. In other, probably the majority of cases, the stimulus for rebuilding is the immediate saving in cost. The degree to which rebuilding is done depends upon the cost and availability of energy, labour and materials. At its simplest, the practicality of such rebuilding depends on the relative costs of new components and of the labour and consumables associated with rebuilding. Traditionally, labour costs have been kept relatively low through the use of power tools in disassembly. Since 1973, however, changes in the costs of energy have been such as to prompt an energy audit of the rebuilding process, to establish whether the re-manufacture of units is energy-efficient as well as cost-effective.

An energy analysis of the routine commercial rebuilding of a car starter motor suggested that the energy of reconstruction was less than 2% of the energy of initial construction. This difference, approaching two orders of magnitude, does not reflect the fact that components rejected as unserviceable for remanufacturing may still be reclaimed as materials, instead of as parts. More than 50% of the total energy required in this instance related to the cleaning and degreasing of yoke assemblies. The most energy-intensive single operation is the heating of the cleaning vats, which consumes some 34% of total direct energy.

11.2 ENERGY CONSIDERATIONS

It is now clear that monetary cost, although still the dominant factor in manufacturing industry, is increasingly being linked with the energy specificity of a process. In the field of energy analysis there are at least two conflicting schools of thought:

(a) Energy analyses are independent of financial analyses; choices may be made on the basis of either or both, and this implies that any material resource can be exploited, given sufficient energy. Energy should thus be treated as a unique and essential resource.

(b) Energy is only one of the primary natural resources that control technology.

The second of these is, initially, the more plausible since not only is it impossible to assign an energy concept of value but it is also clear that other primary inputs, such as capital, labour, land and minerals are also indispensable.[4]

---

[4] Thomas Veach Long II, *Workshop report: Int. Federation of Institutes for Advanced Study, Workshop on energy analysis and economics*, Lidingö, Sweden, (June 22–27 1975).

The concept of the efficient allocation of resources may be defined in numerous ways. It is convenient, here, to give those which relate to efficiency, whether physical or economic:

(a) Maximum physical efficiency is attained when, through the efficient application of energy to a system and the careful husbandry of materials throughout, neither is wasted.

(b) Economic efficiency is attained when given resources of capital, labour and all forms of natural resources are so combined that a higher output of any desired product is possible only at the cost of a reduced output of some other desired product.

Remanufacturing is a process whose spread in developed countries has probably been retarded by the fact that labour is comparatively expensive. Studies published by The World Bank suggest that remanufacturing would be an advantageous activity for developing countries, where a shortage of indigenous manufacturing facilities and of foreign exchange are often accompanied by a readily-available workforce which is well able to acquire the necessary skills. In fact, much *de facto* remanufacturing is already carried out in developing lands, where cars fifty years old are of necessity kept in everyday use by the skills of the indigenous population.

11.3 PROCESS STAGES AND TERMINOLOGY

Remanufacturers may be divided into three types:

(a) The Original Equipment Manufacturer (OEM). The OEM often makes and sells both new and remanufactured versions of its products.

(b) The Independent Remanufacturer. This type purchases unserviceable products and remanufactures them for sale.

(c) The Contract Remanufacturer. This type refurbishes products under contract to a customer who retains ownership of the products.

Remanufacturing commences at the point where a user, through abandonment, sale or trade-in, relinquishes a product to a collection system which forwards them to a remanufacturer. Such unserviceable products are termed 'cores'. In the case of a system maintained by the original equipment manufacturer the collection system usually forms part of the company's distribution channels.

Collected cores are then subjected to:

(a) Disassembly and cleaning
(b) Refurbishing of component parts
(c) Reassembly and testing

Component refurbishing may take a number of forms:[5]

(a) Surface cleaning and preparation are almost standard.
(b) Worn areas are often built up by application of weld metal and then machined to original dimensions.
(c) Holes that wear has made out-of-round may be bored to oversize in order to accept an insert having the correct internal diameter.
(d) Bent shafts are straightened.
(e) Electrical wiring is cleaned and re-insulated.
(f) Precision surfaces are re-ground and scraped.

In each case, the objective is to restore a part to its original condition or, where the design has been found to cause premature failure, to rebuild or replace in such a way that the weakness is eliminated. Thus, it is even conceivable that the remanufactured product may be more serviceable or durable than the original.

Measurement, testing and quality-control methods used in remanufacture are similar to those used in original manufacture but with the one important exception of inspection. All parts must be presumed faulty until proved to be otherwise. Sampling is therefore not permissible and inspection must be on an 100% basis.

11.4 ELIGIBILITY OF PRODUCTS FOR REMANUFACTURE

A wide range of products is remanufactured, but certain common characteristics are clear.[6]

11.4.1 Core value

The unserviceable product must not, during use, be consumed, i.e. there must be a residual core, which must retain sufficient added value in its embodied capital investment, energy and labour, at the end of its useful life, to justify remanufacture. An alternative statement is that the potential worth of the product must exceed its market value as a non-functional product, usually its scrap value. That potential worth is determined by the cost of refurbishing and by the final price that it could, after remanufacture, command in the market place. The equation must, of course, include the cost of disposal; an unserviceable product that incurs disposal costs has a negative market value, which contributes to the argument for remanufacturing.

11.4.2 Market stability

It is essential that a market should exist for the remanufactured product at the price level that must be charged for it. This criterion is largely connected with market stability. The fact that a remanufactured product

[5] Robert T. Lund, *op. cit.*
[6] Lynn Bollinger, Joel Clark, Richard L. Frenkel, Ronald Grand, Richard M. Kutta, Robert T. Lund and Floyd R. Tuler, *Energy Recapture through Remanufacturing. Final Report of Pre-Demonstration Study,* Cambridge MA, Massachusetts Institute of Technology, The Center for Policy Alternatives No.81–13, Jan. 1981.

might be more than ten years old, and its styling therefore dated, might possibly deter anyone other than those in the lowest socio-economic levels of society in developed countries. The question of appearance is, however, sometimes much less important than technological change. Possibly the most spectacular recent change in market conditions relates to the electromechanical calculating machine, which was in common use until, perhaps, the early 1970s. Such machines may well satisfy all other physical and financial criteria for remanufacture, but the electronic calculator has destroyed their market.

Lund has identified four main categories of product which are widely remanufactured in the U.S.A.:

(a) Automotive. This sector contains the largest number of remanufacturers. The parts vary in scope and complexity, from a simple starter solenoid to complete engines.

(b) Industrial equipment. This category includes valves and other hydraulic equipment, metalworking machinery, industrial electromagnets and motors, and a wide variety of others. Since many of the products were initially custom-built they tend to be remanufactured individually.

(c) Commercial products include office machinery, vending machines, and communications equipment, also remanufactured individually.

(d) Residential products. Remanufacturers are, or were, relatively uncommon in the U.S.A. This was attributed to:

  • A highly disaggregated market for used products. The supply of cores will be similarly disaggregated.
  • A long product life, which tends to emphasize the effects of technological obsolescence.
  • High transport costs.
  • Customer prejudice against rebuilt products.

Remanufacturing in the domestic sector tends to be concentrated in the area of power tools, garden and leisure equipment, and small appliances. It is evident that, in these categories, the effects of styling are likely to be slight. Loudspeaker drive units are examples of products whose visibility is low, and whose appearance is virtually irrelevant. Their performance, however, may be improved significantly and inexpensively by change of cone or magnet material.

The largest remanufacturer in the U.S.A. is evidently the Department of Defense, which has a continuous programme of remanufacturing designed to maintain items ranging from small arms to battleships — several examples of the latter, dating from World War II, have so far been recommissioned with modern weapons systems — with a view to maintaining optimum operating efficiency and to incorporating technological advances at much lower cost than building from new.

113

11.4.3   Product design

Product design plays a key part in the criteria for selection of products as suitable for remanufacture.

What may be termed 'design maturity' is an important factor in remanufacturing. Remanufacturers avoid those products whose design is undergoing rapid change, in either design or materials. The products favoured for remanufacture are those with a slow rate of design change from year to year. As in most manufacturing processes, a steady supply of raw material of consistent quality is very much to be desired. Thus, there must be a high ratio of technological life to product life, to ensure that the product does not become obsolete before it wears out.

The product must be standardised and made with interchangeable parts which have been factory-assembled. It must be capable of disassembly. In metal components, welded, soldered or swaged joints, while less easy to take apart than bolts, can be separated. Plastic bodies joined by rivets are a more difficult and vulnerable proposition.

The component parts must be capable of repair, refurbishment or economic replacement. The final product must be capable of reproducing the original performance.

The product should also have a low ratio of undifferentiated material to total material. Undifferentiated materials are defined as those which are not formed or worked for the application in hand, but which could be used for other purposes. This is a measure of the likely amount of added value likely to be incorporated in the product. A length of chain or a dismantleable brick wall would each have a high proportion of undifferentiated material.

11.5   BARRIERS TO REMANUFACTURING

A major obstacle is prejudice toward a remanufactured product. Market attitudes are quixotic, sometimes imponderable, and often intractable, especially in sectors in which the purchaser is not expert or where newness is considered as important as utility. However, this class of obstacle is concerned less with remanufacturing per se than with marketing.

Rapid or frequent design changes in manufactured products may inhibit any interest in entering into remanufacturing them. A major difficulty associated with such changes is that parts for newer models tend to be incompatible with older units.

Remanufacturing may be inhibited by lack of replacement components. This arises, for example, if an OEM refuses to sell parts for remanufacturing, or if it refuses to divulge specification information that would permit the manufacture or testing of facsimile or generic parts.

The process of remanufacturing is seen as a most desirable operation which should be encouraged in contexts where an unserviceable product will, after rebuilding, still deliver an acceptable standard of safety.

12

THE CAPACITY OF THE SECONDARY INDUSTRIES TO ABSORB RECOVERED MATERIALS

12.1 INTRODUCTION

Unless their recovery is desirable on environmental grounds there is no point in recovering materials for which there is no market. Therefore, the viability or otherwise of materials recovery depends entirely on the capacity of existing or potential markets to absorb them.

The prices of many secondary materials are volatile in that they are subject to fluctuating demand in the marketplace. Since, in many cases, secondary material provides only a small fraction of total materials requirements, it is usually the first to be given up should production need to be reduced. For this reason, statements based on price may not remain valid for more than a few weeks or months in the face of external changes in the general level of activity in the economy.

The ability of industry to absorb secondary materials may be discussed in terms of quantity, quality and price. High-grade material is generally saleable, since it may be used in all classes of application. Low-grade material may find only limited areas of application. Both grades must be competitive in price.

12.2 AUTOMOTIVE SCRAP

The importance of automotive scrap to the supply of secondary materials is clear. Changes in vehicle construction and composition must therefore be of great significance to the recovery industry, and will serve as an illustration.

By contrast with the U.S.A., there appears to be no reliable published information on the materials composition of automobiles produced in the different countries of Europe. In the last twenty years, the number of independent car manufacturers has diminished, and most of the remainder have production units in more than one country. Many models which are nominally the product of one country incorporate several major components, each made in a different land. The emergence of the so-called 'World Car', designed to be sold in basically similar form but with minor modifications in all the leading world markets, provides further stimulus toward standardisation. Cooperative deals, such as that between BL and Honda will further reduce the number of independent car-makers, who may well number as few as four by 1990.[1] Even so, since the European automotive industry is based in several countries it is unlikely to approach the relative homogeneity of its U.S. counterpart.

A comparison has been made of the various published evaluations of the likely composition of average European cars.[2] The projections vary widely and there can be little confidence in data that are so widely scattered. Moreover, as is well known, projections can seriously be invalidated by unforeseen political developments that may change the availability of materials. However, on the basis of the predictions made by Bashford,[3] Whalley has proposed scenarios that might be used to calculate possible future materials availability from European cars.[4]

The discussions which follow are intended not as a comprehensive analysis of metallic scrap flows but to put the quantities of scrap potentially available from automobiles into the context of the total quantities used in the ferrous and aluminium industries.

12.3   FERROUS METALS

On the basis of a 75% ferrous content in cars, Europool has presented data that suggest that the quantities of iron and steel theoretically recoverable from them in 1975 ranged from 2·5% (Belgium and Luxembourg taken together) to 9% each (France and Netherlands) of the scrap consumption of the steel industries of the EEC.[5]

The data presented by Bashford and by Whalley facilitate an analysis of the effects of auto-derived ferrous metals on the steel industry.[6]

Steel production in Europe (Benelux, France, Federal Republic of Germany, Italy, Portugal, Spain, Sweden and the U.K.) for 1982 was 127 million tonnes, compared with 146 million tonnes in 1979. That period of programmed capacity reductions may by now have finished; however, on the basis of present data there seems no justification for assuming any increase in steelmaking capacity. An analysis by process route, assuming scrap rates of 45% for the remaining open hearth furnaces, 95% for the electric arc furnace, and 30% for the basic oxygen furnace, suggests that European steelmaking scrap requirements were approximately 63 million tonnes. Assuming that 50% of that would be purchased scrap, likely demand is calculated as approximately 31.5 million tonnes.

[1] N.A. Waterman, *Materials in the family car — an investigation of the possibilities for the substitution of indigenous materials for imported raw materials in the U.K. automotive industry,* unpublished report to The Materials Forum, 1982.

[2] Michael E. Henstock, The European Picture, chapter in *The Impacts of Material Substitution on the Recyclability of Automobiles,* New York, The American Society of Mechanical Engineers, 1984.

[3] G.D. Bashford, Design for energy saving, *Future Metal Strategy,* London, The Metals Society, 1980.

[4] L. Whalley, The effect of changing patterns of materials composition on the recycling of cars and domestic appliances in the U.K. automotive industry, *Conservation & Recycling,* Vol.5, No.2/3, 1982, pp.93–106.

[5] Europool, *The Disposal and Recycling of Scrap Metal from Cars and Large Domestic Appliances,* London, Graham & Trotman Limited, 1978.

[6] Henstock, *op cit.*

Assumptions:

(a) A ten-year lifetime for vehicles entering service in 1979
(b) A composition equivalent to Whalley's Scenario V (steel 75%, aluminium 3%)
(c) No increase in steelmaking capacity 1982–1989
(d) No change in the balance of steel manufacturing route

Under such assumptions, the 11·46 million vehicles built in Europe in 1979[7] and (notionally) scrapped in 1989 could supply some 9·45 million tonnes of ferrous scrap, i.e. 30% of purchased steel scrap demand, assuming that quality standards made the scrap acceptable. Under Scenarios III and IV, (Steel 55%, aluminium 5%; and steel 55%, aluminium 20%, respectively, of a vehicle some 27% lighter than under Scenario V), some 5·04 million tonnes of steel scrap would be available, i.e. 16% of purchased scrap demand.

A similar analysis of an assumed 12.5 million vehicles built in 1985[8] and scrapped in 1995 suggests that they could, under Scenario V, supply 33% of purchased steel scrap requirements. Scenarios III and IV suggest an availability of 17% of requirements. Such small rises, relative to 1979, must clearly be deemed insignificant given the obvious uncertainties in the assumptions made for composition and numbers of vehicles, and in the future size of the steel industry.

If the automotive steel scrap were in baled form, whose undesirability was demonstrated in Chapter Four, it is possible that supply might exceed demand. Given, however, the move towards shredded scrap, which contains less copper than baled scrap, it is judged likely that steel derived from automobiles will not, by 1995, form an unacceptably large fraction of total purchased scrap requirement unless the quality of the scrap deteriorates to the point of unacceptability, or unless that requirement actually declines. In this context it should be noted that the scrap rate in the BOF of many Japanese integrated steel mills is as low as 5%, and that this is almost met by scrap generated in the continuous casting process. The extension of such practices to Europe could, unless counterbalanced by increased scrap consumption in electric furnaces, have a serious effect on demand for purchased scrap. The reduction of steel use in vehicles will help to mitigate such oversupply. Another point of relevance is possible protectionism in the face of worldwide overcapacity in steel production.

If the steel industries of Europe were unable to absorb it, automotive shredded scrap could, on grounds of quality but possibly not of price, substitute for lower-grade raw materials currently used by iron foundries.

12.4   LOW-GRADE FERROUS SCRAP

Table 12.1 shows the potential of various market sectors for absorbing low-grade scrap.

[7] Waterman, *op. cit.*
[8] Waterman, *op. cit.*

Table 12.1
Steel product suitability for inclusion of low-grade scrap[9]

Product	Typical % of shipments	Suitability of low-grade scrap as ingredient
Reinforcing bars	5·4	Excellent
Selected hot-rolled light sections	6·7	Excellent
Selected wire rods	1·8	Very good
Selected rail accessories	0·5	Very good
Selected plates	8·6	Good
Heavy structural shapes	7·5	Fair
Steel piling	0·5	Fair
Hot-rolled strip	1·4	Marginal
Hot-rolled sheet	13·6	Marginal
All other products	54·0	Generally unsuitable

In plants producing a large variety of products, including high-specification items, low-grade scrap would be unattractive even at low prices, since it cannot be moved from one application to another and also generates home scrap of low quality. The total market for reinforcing bars and light sections could absorb all the recoverable low-grade ferrous scrap if it were all processed by minimills. The large integrated steel producers, who share the market, are reluctant to use low-grade scrap. However, it should be noted (Section 4.4.4. and Table 4.2) that 78% of the output of a typical steelworks is able to tolerate 0·30% Cu and 0·05% Sn.

The other large market for ferrous scrap is in foundry products. The foundry industry is, after steelmaking, the largest market for ferrous scrap, and it absorbs some 20% of U.S.A. supply. Its products are:

(a) grey and ductile iron,
(b) malleable iron, and
(c) steel castings.

Some 85% of industry output falls into category 1.

Foundry products are generally much more tolerant of tramp elements than are the products of steel mills; grey iron and pearlitic ductile iron may each contain up to 1·50% Cu and, respectively, 0·15% and 0·10% Sn.

Elements other than iron are important for their effects on the structure of ferrous foundry products; heat treatment is also critical. Two varieties of cast iron are produced by controlling the rate of solidification:

(a) grey iron, containing most of its carbon in the form of graphite flakes, which confer the characteristic colour on a fracture surface,

(b) white iron, which retains most of its carbon as iron carbide (cementite), Fe_3C.

Residuals in foundry products have complex effects, not only in their role as solid-solution hardeners but also in their effect on the form in which the carbon appears. Graphite formation is promoted by aluminium, copper, nickel, silicon, and titanium, and retarded by chromium, manganese, molybdenum and vanadium.

Sulphur may retard graphitisation and can cause difficulties during solidification. Phosphorus may have adverse effects on the strength of a casting, but by increasing melt fluidity is beneficial in thin-wall castings.

Up to 2% Cu may be present in grey cast iron; certain alloy irons may contain up to 7% Cu and are used for their corrosion resistance. The effects of copper are otherwise similar to those of nickel.

The additional use of post-consumer tinplate scrap, especially in its incinerated state, has been viewed as potentially destabilising.[10] The matter stimulated much research during the 1970s, to establish with greater precision the effects of copper, tin and other contaminants on the properties of ductile iron.[11, 12]

The proportion of scrap used in cast iron cupola burdens in the U.S.A. increased from 64% in 1950 to 90% in 1975 and to 95% in 1985.[13] Increasingly stringent environmental legislation seems likely to cause a shift in productive capacity towards the electric furnace, which operates almost entirely on scrap. This development may offer still greater opportunities for scrap consumption. Maximum penetration will, however, depend on the following:

(a) The production of physically clean scrap, whose use will not produce air pollution, and
(b) the development by the scrap industry of a so-called 'specification ingot', melted to a known composition, size and specific gravity.

12.5 ALUMINIUM

It has been seen that aluminium, if not already so, is likely to be the most important component of the non-ferrous fraction of shredder output, as 1980 and later model cars reach the scrapyards.

Total U.S. consumption of aluminium in 1985 was 6·162 million tonnes, of which 1·762 million tonnes (29%) was derived from scrap. Some 850 000 tonnes, 48% of total secondary metal, came from old

[9] Anon, *First Report to Congress: Resource Recovery and Source Reduction*, Washington, D.C., US Environmental Protection Agency (SW-118), 1974.
[10] Michael B. Bever, The recycling of metals — I. Ferrous metals, *Conservation & Recycling*, **1**, (1), pp.55–59, 1976.
[11] L.A. Neumeier, B.A. Betts and D.H. Desy, *AFS Transactions*, **82**, (74–59), pp.131–138, 1974.
[12] L.A. Neumeier and B.A. Betts, Ductile iron containing tin, copper and other contaminants, *AFS Transactions*, **84**, (76–94), pp.265–280, 1976.
[13] Staff, Inst. of Scrap Iron and Steel, Inc., *Facts, 1985 Yearbook*, 43rd Edition, Washington, Inst. Scrap Iron and Steel Inc., (ISIS), 1986.

scrap. This showed a distinct improvement over 1975, when only 306 000 tonnes, or 27%, of secondary aluminium came from old scrap.[14]

The difference may well be associated with increased efficiency of recovery of used aluminium beverage containers. Much aluminium is used in packaging and is probably buried in that form. The quantities of old aluminium scrap potentially available in cars in 1990 are likely to be much increased relative to those suggested by Dean and Sterner for 1969. Utilisation of this fraction would clearly require a change in industry structure, with a greater use of old relative to new scrap than has hitherto been the case.

In Europe, the 12·5 million vehicles forecast to leave service in 1995, of a representative 800 kg weight and containing 5% Al, would contain approximately 0·5 million tonnes of secondary aluminium, and this may be assumed to be in the ratio 3:1 cast to wrought.

The principal product of the secondary aluminium industry is casting alloys; it has been estimated that in the U.S.A. 90% of secondary aluminium goes into castings, with most going directly to secondary smelters. There, the heterogeneous mixture, of unknown composition and containing numerous impurities, is smelted into a product of known composition but suitable only for castings. However, casting alloys comprise only about 20% of total aluminium consumption. The remainder is mostly in wrought form.

The secondary aluminium industry does not compete with the primary producers so far as wrought metal is concerned, because secondary aluminium is generally of lower purity than is primary. However, the raw material for the secondary producers, i.e. obsolete (old) and prompt industrial (new) scrap, consists largely of wrought alloys, since these are used in aluminium products in the ratio of about 4:1 wrought:cast. Thus, primary material is used principally in wrought products, yielding both new and old scrap which is of controlled purity. This scrap then passes mainly to secondary metal producers whose products are saleable only as casting alloys. Some new scrap can, if carefully segregated by type, command from the primary producers a price that is too high for the secondary smelters to match.

It may thus be seen that an increase in secondary metal production relative to primary implies an increased production of and market for casting alloys, relative to wrought, or a need for secondary producers to enter the wrought alloy market, a course that would require better identification and segregation procedures.

The automotive industry differs from other sectors of the economy in that at present most of its consumption is of casting alloys. The search for lighter weight will, though, stimulate increased use of aluminium in structural members such as sub-frames, bumpers, and sheet-metal body parts. These will require wrought aluminium alloys; it has been estimated

[14] *Metallstatistik 1975–1985*, 73rd Ed., Frankfurt am Main, Metallgesellschaft Aktiengesellschaft, 1986.

that of the projected increase in aluminium use in automobiles between 1980 and 1990, two-thirds will be wrought.[15]

The problems in using scrap for producing wrought alloys lie in their relatively stringent specification. Alloys for die, for permanent mould, and for sand-castings are typically permitted up to 1% Fe and 0·5% Zn; some widely used alloys equivalent to the U.S. Type 380 even permit up to 3% Zn. By contrast, wrought alloys may be limited to 0·25% maximum Zn, and 0·5–0·7% Fe.

No data are available for the production of aluminium castings in several European countries. However, these are principally those whose consumption of aluminium is relatively modest; Austria, Netherlands, Norway, Spain, Sweden and Switzerland together account for only 20% of total European consumption, with Spain alone consuming one quarter of this. Assuming that aluminium castings are produced in these countries in quantities relating to overall aluminium consumption, total castings production in Europe in 1985 may be estimated as about 1·13 million tonnes.

A possible increase in the aluminium content of cars, from the assumed 5% to 20% or more, can only exacerbate the problems of absorbing the metal when it is reclaimed. There is some slender evidence that in the years following the major oil crisis of 1973/74, the use of secondary metal increased relative to primary in Europe.[16] However, over the entire period 1968–1985, the ratio of total (primary plus secondary) aluminium consumption to aluminium recovered from scrap has never varied by more than 5% from an average of about 3·9:1.[17]

12.5.1 Impact of recycling on the aluminium castings industry

A comprehensive analysis of the processes involved in the remelting of scrap aluminium to produce secondary metal is outside the scope of this discussion. The question of compositional ranges and limits and the effects of trace elements are a subject of enormous complexity. All that will be attempted here is an illustration of some of the problems involved.

At various times, concern has been expressed over the ability of the primary and secondary aluminium industries to meet the projected increased demand for castings in automotive applications. Increased scrap recycling was thought likely to provide the source of metal for this expansion of the castings industry in the U.S., in the face of possible or probable shortages of primary metal and in view of the cost and energy advantages of using secondary metal.[18]

---

[15] Roig *et al.*, *op. cit.*
[16] Henstock, *loc. cit.*
[17] *Metallstatistik 1975–1985*, 73rd edition.
[18] E. L. Rooy, Aluminum Scrap Recycling and Its Impact on the Metal Castings Industry, AFS Transactions, 85–179, pp.935–8, 1985.

Three developing scrap streams were thought adequate to satisfy the most optimistic projections for growth in the castings industry. These were:

(a) directly-reclaimed beverage cans (UBC) and can scrap,
(b) secondary metal (mostly beverage cans, UBC) from municipal solid waste (municipal), and
(c) aluminium from auto-shredder residues (auto).

Many problems, however, lie in compositional compatibility. It is normally assumed that, whereas wrought grades are usually intolerant of contamination by recovered castings, the reverse was not true, and that castings would accept all, or almost all, classes of wrought material. This is, however, untrue. Table 12.2 shows the average composition of aluminium scrap derived from the three sources given above.

Table 12.2
Typical aluminium scrap chemistry by source[19] (E. L. Rooy *loc. cit.*)

Source	Element (%)					
	Si	Fe	Cu	Mn	Mg	Zn
Municipal	0·8	0·5	2·4	0·6	1·7	2·0
Automotive	5·0	0·8	1/3	0·3	0·1	0·6
UBC	0·2	0·6	0·15	0·9	1·1/1·3	

Separation from municipal solid waste is still not widely practised. Rooy estimated that less than 1% of the total metal contained in municipal solid waste is actually reclaimed for recycling. The UBC, in any event, presents problems in its high manganese content which, in the die casting industry, contributes to sludge formation and whose use is constrained by restrictive specification limits in gravity casting. Recent developments have identified compositions in which manganese can not only be tolerated, but might even be advantageous.

Metal scrap will generally be employed in those applications which offer the greatest potential return. Used aluminium beverage cans (UBC), assuming that they can be segregated from contamination, provide stock for more beverage cans. Thus the primary industry is much involved with the development of systems for the recovery of the UBC. Similarly, the most logical end use for aluminium scrap recovered from vehicles is in the foundry alloys from which the recovered metal is derived.

Scrap segregated from municipal solid waste is relatively contaminated and has low usability, except at high dilution rates, for even the broadest alloy compositions.

[19] E.L. Rooy, *loc. cit.*

Automotive scrap is more flexible. However, a very wide range of alloys is used in automotive applications.[20] The compositions of the wrought alloys used for bumpers, body sheet and other applications are often incompatible with foundry compositions. The Alcoa development of Type 5182 aluminium-magnesium-silicon alloys, for body-sheet application in competition with the European 2036, would ease the recyclability of automotive aluminium scrap in compositions for foundry use, as shown in Table 12.3.

Table 12.3
Usability of aluminium scrap in foundry alloys[21] (E. L. Rooy, *loc. cit.*)

Source	Usability	
	319·1 grade (%)	A380·1 grade (%)
Municipal	50	100
Automotive	100	100
UBC	50	50

Remelting shops must deal with a wide range of incoming scrap sources, from which they generally make a more-or-less wide range of products. Some scrap sources may be captive, obtained either from sister companies or from scrap produced by their own customers.

Typically, foundries are offered and may accept all kinds of scrap, from turnings to old domestic appliances, cooking pots, off-cuts, castings and some secondary ingot. Old castings are lumped together in one type of scrap. Some companies segregate by end-use alloy, e.g. they know which alloy is used for pistons, blocks and other applications and can sort accordingly. Foundries which do not sort incoming arisings may encounter difficulties with iron, magnesium or zinc. Copper and silicon are generally not a problem. However, incoming pressure die-casting alloys will bring with them up to 1% Fe and 3% Zn, and these can be handled economically only by those plants which themselves produce such alloys.

The problems were analysed at length by Siebert as long ago as 1970, and much of that analysis is still valid.[22] Melts have sometimes to be rejected because they cannot be brought into specification within the size of the available furnace. However, alloy production is pre-planned, and typically only perhaps 1% of melts is so far outside specification as to be unsalvageable without being allowed to solidify and held for future use.

---

[20] Joseph J. Tribendis, C. Norman Cochran and Ronald H.G. McLure, Aluminum Alloys, Chapter in *The Impacts of Material Substitution on the Recyclability of Automobiles*, New York, The American Society of Mechanical Engineers, 1984, pp.119–145.

[21] E. L. Rooy, *loc. cit.*

[22] Donald L. Siebert, *Impact of technology on the commercial secondary aluminum industry*, Washington, D.C., U.S.Department of the Interior, Bureau of Mines, IC 8445, 1970.

The most frequent cause of rejection is iron. All melts will eventually be reassimilated but with obvious losses in drosses and energy.

Techniques exist for separating aluminium from mixed scrap, but there are as yet no efficient methods of separating wrought aluminium from cast. An interesting development exploits differences in their mechanical properties at high temperatures.[23, 24]

A major problem facing many secondary smelters is the air pollution which arises from the combustion of contaminants such as oil, paint and non-metallic attachments such as polymers.

The question of scrap reassimilation is a worldwide one. In the USSR, aluminium scrap and wastes are used for the direct production of casting alloys.[25]

12.5.2 Specialised aluminium alloys

A significant recent development in the field of aluminium alloys involves the use of lithium as an alloying constituent. The concentrations of lithium used are small, typically of the order of a few thousand ppm. However, aluminium–lithium alloys differ from other aluminium alloys in respect of the high value of the alloying additions, the high reactivity of the material, and the absence of lithium from the specification of registered alloys other than those specifically described as aluminium–lithium compositions.

The value of lithium is high, some £50 000/tonne. Therefore, the lithium costs more than the aluminium for each of the current generation of Al–Li alloys and offers financial scope for recovery in ways other than by simple recycling into aluminium–silicon foundry alloys.[26]

New scrap processing and treatment technology is likely to be needed. In the short term, while market quantities of Al–Li alloy scrap are small, only the aluminium content is likely to be recovered, after elimination of the lithium. In the medium term, perhaps 3–5 years, new processes are likely to be developed for lithium recovery, with subsequent recycling into aircraft-grade metal.

It is estimated that by 1995 two-thirds of the aluminium used in the aircraft industry may be Al–Li alloys, and may amount to perhaps 100 000 tonnes per annum. Lithium is also highly reactive; both it and molten Al–Li will react with humid atmospheres and are aggressive towards refractories. Hence, the particular problems presented to the secondary aluminium industry by these alloys will need to be addressed. They exemplify the difficulties associated with new materials

[23] F. Ambrose, R. D. Brown, Jr., D. Montagna and H. V. Makar, Hot-crush technique for separation of cast- and wrought-aluminum alloy scrap, *Conservation & Recycling*, **6,** (1/2), 1983.

[24] R. D. Brown, Jr., F. Ambrose and D. Montagna, Separation of cast and wrought aluminum alloys by thermomechanical processing, Washington, D.C., United States Department of the Interior, Bureau of Mines, RI 8960, 1985.

[25] V. A. Popov, *Tsvetn. Met. (Moscow)*, (6), pp.64–5, 1987.

[26] W. R. Wilson, J. Wort, E. P. Short and C. F. Pygall, *J. Phys. Colloq.*, (C3), C3–75/C3–83, 1987.

12.6   COPPER

The refining processes that may be carried out on copper are capable of producing secondary metal indistinguishable from primary. Therefore there need be little doubt that secondary copper produced by a full refining route can be reassimilated into the materials stream. However, the cost of this operation is such as to make it advantageous to segregate scrap, so that part of the refining operation may be avoided.

The metallurgical effects of impurities in recycled copper alloys have recently been reviewed from the point of view of hot shortness and of undesirable phase transformations.[27]

Several techniques are used routinely in the copper industry to identify scrap for effective segregation. These are based on object recognition, colour, density, magnetic properties, and chemical spot tests.

12.6.1   Impurities in copper alloys

Hot shortness in wrought alloys, i.e. a loss of ductility at temperatures greater than the recrystallisation temperature, is caused by such low melting-point contaminants as antimony, bismuth, and lead. Cold shortness, or ductility loss at temperatures below the recrystallisation temperature, is associated with very low concentrations of antimony, bismuth, boron and tellurium.

As a rule, castings can tolerate greater concentrations of impurity elements than can wrought, and copper and its alloys are no exception. However, there is no general rule about permissible levels of contamination; for example, concentrations of antimony and lead which may be tolerated in some alloys are intolerable in others.

As in the case of aluminium, the question of permissible residual levels is one of great complexity. Generally, antimony, bismuth, and lead are the most damaging elements in wrought alloys. Their presence can severely impair the recyclability of copper alloys.[28]

It is axiomatic that any increase in recycling level of any material must be accompanied by the existence of suitable markets for the products.

[27] Harry V. Makar and William D. Riley, *Metallurgical effects of impurities in recycled copper alloys*, Washington, D.C., United States Department of the Interior, Bureau of Mines, IC 9033, 1985.
[28] Harry V. Makar and William D. Riley, *op. cit.*

INDEX